AN ILLUSTRATED HISTORY OF

MILITARY HELICOPTERS

AN ILLUSTRATED HISTORY OF
MILITARY HELICOPTERS

FROM THE FIRST TYPES DEPLOYED IN WORLD WAR II
TO THE SPECIALIZED AIRCRAFT IN SERVICE TODAY,
SHOWN IN OVER 200 PHOTOGRAPHS

FRANCIS CROSBY

southwater

This edition is published by Southwater,
an imprint of Anness Publishing Ltd,
108 Great Russell Street,
London WC1B 3NA;
info@anness.com

www.southwaterbooks.com; www.annesspublishing.com; twitter: @Anness_Books

A CIP catalogue record for this book is available from the British Library.

Publisher: Joanna Lorenz
Senior Editor: Felicity Forster
Designer: Nigel Pell
Production Controller: Rosanna Anness

Previously published as part of a larger volume, *The World Encyclopedia of Military Helicopters*

ACKNOWLEDGEMENTS
Picture research for this book was carried out by Jasper Spencer-Smith, who has selected images from JSS Collection and the following (key: l=left, r=right, t=top, b=bottom, m=middle): UK MoD Crown Copyright 2010: 4bl; 5br; 7b; 38–9; 42br; 57tl; 60t; 67bl, 67br; 68b; 69tr; 70t; 85t, 85b; 88bl; 95. Every effort has been made to acknowledge photographs correctly, however we apologize for any unintentional omissions, which will be corrected in future editions.

PAGE 1: **Mil Mi-8 (NATO identifier Hip).**
PAGE 2: **Bravo November.**
PAGE 3: **Bell UH-1A Iroquois Huey.**
BELOW LEFT: **Westland Sea King.**
BELOW RIGHT: **AgustaWestland EH-101 Merlin.**
OPPOSITE LEFT: **Sikorsky S-51.**
OPPOSITE RIGHT: **Boeing CH-47 Chinook.**

Contents

Introduction

Although the word "helicopter" is derived from the ancient Greek *heli* (twisted, curved) and *pteron* (wing), the word we recognize today was not suggested until the 1860s in France, as *hélicoptère*. Both cultures are, however, pre-dated by the idea of flying in a way that we now associate with helicopters and a few fixed-wing aircraft. Successful helicopter designs are, compared to fixed-wing aircraft, relatively recent innovations. While thousands of monoplanes, biplanes, triplanes, bombers, scouts and fighter aircraft were being produced in World War I, helicopters were not developed in a concerted manner, and then only in a limited manner, until World War II.

From novel, unstable and frightening machines, helicopters have been developed to become incredibly sophisticated flying machines that can fly forward, backward, sideways and, of course, hover. If the military were initially slow to appreciate the potential of helicopters, they have been making up for that short-sightedness ever since. Helicopters have been developed for an incredible variety of military roles, including minesweeping, reconnaissance, rescue, casualty evacuation, gunship, tank-busting, anti-submarine warfare and heavy lifting. The helicopter has also been armed with guns and missiles, and even equipped to carry nuclear depth charges. The military troop-carrying role is among its most important, and the machine's unique capability enables military commanders to insert combat-ready troops into battle.

The civilian uses of the helicopter are no less impressive, and include firefighting, rescue, crop-dusting, law enforcement and of course being the transport of choice for the wealthy and famous. According to the American Helicopter Society (AHS), there are over 45,000 helicopters operating in the world today, and over three million lives have been saved by these aircraft in both peacetime and wartime operations since the first rescue at sea in 1944.

ABOVE: **Britain's helicopter designers developed a range of unique designs such as the Saunders-Roe Skeeter, pictured here, but the industry was ultimately dominated by licence-built versions of US designs and, later, types designed by international consortia.**

LEFT: **The Bell 47 H-13B Sioux, with its distinctive skeletal tailbone and heavily glazed round cockpit, was a classic military helicopter that was in manufacture for 27 years.**

While early designers may have grasped the broad concept of rotary flight, the power to get this kind of machine airborne simply did not exist. As engine technology developed, so did the ability to get designs from the drawing board into the air. Piston engine technology was accelerated during World War I, as was jet turbine technology during World War II.

Early military helicopters were powered by piston engines and gave sterling service, but the introduction of turboshaft engines dramatically boosted the helicopter's performance in terms of endurance and speed.

Better knowledge of aerodynamics, technical developments, computer-aided design and the creation of lightweight composite materials have led to a huge range of advances in helicopter design, which in turn have boosted all-round performance and lifting capability. Avionics and the incredible array of equipment and weapons that can be carried by a helicopter has made the type among the most expensive in the military inventory.

The helicopter is a versatile and vital asset in military inventories around the world, and plays a truly unique part in aviation that at the time of the Wright brothers' historic flight could only been predicted by a few true visionaries.

This book explores the history and evolution of these fascinating aircraft, from the earliest pioneers to the powerful machines of today. It features biographies of the most important innovators in the history of military helicopter development, as well as cutaway diagrams showing the interior structures of various helicopter examples. Key technical features are discussed in detail, such as the cockpit, rotor systems, engines, Night Vision Technology (NVT), tilt-wing and tilt-rotor and Vertical Unmanned Aircraft Systems (VUAS), and a useful glossary of terms and abbreviations is included at the back of the book.

RIGHT: **Helicopters developed for military use, including this Mil Mi-14, have also served in a range of humanitarian roles, including rescue from the sea.**
BELOW: **The French-designed Aérospatiale SA 330 Puma has been widely exported and has been in Royal Air Force service since the early 1970s. The twin-engined Puma has been developed into the Super Puma built by Eurocopter.**

U.S. AIR FORCE
59-1567
91567
DANGER

Pioneers and early development

After early experiments and thanks to the dedication of gifted engineers and designers, rotary craft concepts were refined and developed into a variety of trailblazing designs. War has always proved to be a fertile ground for technical advancement and innovation, and so it was with World War II. The Royal Air Force used autogyros to calibrate the radar that gave the British a vital edge while the threat of German invasion loomed in 1940, and the Royal Navy were operating US-built helicopters by the end of the conflict.

Meanwhile, the German applications of helicopter technology during World War II are often overlooked, but included the large and impressive Focke-Achgelis Fa 223. Although it was only built in small numbers, the Fa 223 proved its effectiveness as an all-weather transport aircraft, and even recovered a crashed Fa 223 using the now universal underslung cargo net technique. While the German Navy used rotor gliders operating from U-boats as early warning aircraft, the Luftwaffe's little-known Flettner Fl 282 was also operated from German ships in addition to its land-based role as a front-line artillery-spotting aircraft.

US forces used Sikorsky's R-4 in front-line service towards the end of World War II, and in April 1944 the type was used in Burma to carry out the first-ever combat rescue mission by helicopter. The age of the military helicopter had arrived.

LEFT: **The Kaman H-43 Huskie was flown on more rescue missions during the Vietnam War than any other type of helicopter. Although it was fitted with an unusual twin rotor system, the type had conventional collective and cyclic flight controls.**

ABOVE: **Leonardo da Vinci's 15th-century design was a flight of fancy that could not even fly in model form.** RIGHT: **In 1854, Sir George Cayley produced his "Aerial Top" to show the lifting power of the propeller.**

Helicopter visionaries

In common with many fields of human endeavour, the creation and development of what we now know as the helicopter began centuries ago as a concept in the minds of visionaries. These visionaries, apparently inspired by nature, had a notion of a craft somehow powered to achieve flight. While some sought to emulate the flapping wings of birds, others were inspired by the simple rotating seeds of trees such as the sycamore, which could fly for some distance with the correct wind conditions – as long as the seed continued to turn and wind generated lift, the seed could stay aloft.

Around 400BC, the Chinese are known to have been making special spinning tops consisting of slightly angled feathers fixed to the end of a stick which, when rapidly spun between the hands, generated lift that could carry the top upwards for free flight. These were just toys, but around AD 400, again from China, we have the first written reference of the notion of rotary wing aviation. A book entitled *Pao Phu Tau* refers to a person called "The Master" who describes flying cars made of wood from the inner part of the Jujube tree, powered by leather straps fastened to returning blades that set the machine in motion. This machine, however fanciful or theoretical, is close to what we understand today to be a helicopter.

For the first signs of an appreciation of helicopter principles in the Western world, we have to look to Ancient Greece and Archimedes, the inventor, mathematician and physicist who, in the 2nd century BC, perfected his rotating screw for use as a water pump. As the screw was rotated inside a cylinder, the water in front moved, and simultaneously the water resisted and pushed back. This resistance principle also applied to the movement of a screw through air, which has fluid properties, to produce lift.

Many mistakenly believe it was Leonardo da Vinci who invented the first helicopter. He stated in his *Codex Atlanticus* that he had discovered a screw-shaped device made of iron, wire and starched linen, around 4m/13ft across, that would rise in the air if turned quickly by a team of four men. His theory for compressing the air for lift was in essence similar to that of today's helicopters. Genius though he was, the intriguing "Helical Air Screw" that Leonardo drew around 1483 (but not published for another 300 years) was, however, nothing more than a concept. Leonardo's plan to use just muscle power to revolve the rotor would never have been sufficient to operate a helicopter successfully, and there was no provision for dealing with the torque created by the turning blades. A scale model made in the early 21st century, even with the advantage of a lower weight than the full-size machine, could not raise itself in flight. Numerous experiments, inventions, technical developments and advances over the centuries following Leonardo did, however, ease the way for the development of the helicopter and vertical flight. The ability to produce robust precision mechanical parts came with the Industrial Revolution, and after that huge technological leap, the production of the helicopter as we know it today was inevitable.

In 1754, the "Father of Russian Science" Mikhail Lomonosov designed a machine with coaxial rotors to take meteorological instruments skywards to take readings. Perhaps inspired by the same principles as the Chinese feather toy of 2,000 years earlier, Lomonosov's model used two propellers rotating in opposite directions on the same level. Lomonosov demonstrated a model powered by a clock mechanism to the Russian Academy of Sciences in July 1754, but it is unsure whether his device flew unaided.

In 1768, French mathematician J. P. Paucton proposed a man-powered machine called a Pterophere, with two airscrews – one to provide lift for the machine in flight, and the second to provide forward propulsion. This machine was still based on the notion of "screws" boring through the air like a screw through wood. Then, in 1783, again in France, Launoy took the Chinese top concept that had been examined a few years earlier by Lomonosov, and produced a model consisting of two sets of rotors made of turkey feathers that rotated in opposite directions. This was demonstrated in 1784 in front of the Academy of Sciences, and succeeded in achieving free flight.

The last significant helicopter visionary was the British engineer Sir George Cayley, who famously built the first practical aeroplane in 1853, but had first sketched a twin-rotor helicopter in 1792. Cayley spent most of his life experimenting with flying machines, and carried out the first serious, experimentally based aeronautical research. In 1843, Cayley designed his "Aerial Carriage" which had four "rotors" arranged coaxially in pairs. However, Cayley was unable to find any steam engines that were light enough to help the design leave the drawing board and, hopefully, the ground.

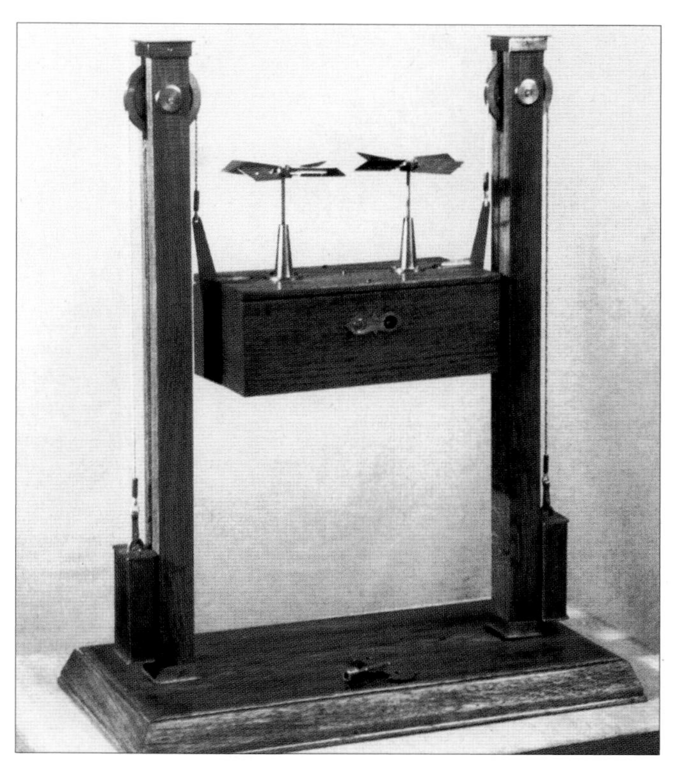

ABOVE: **Mikhail Lomonosov demonstrated this machine in 1754 to the Russian Academy of Science, to prove the lifting capacity of the propeller.**

Steam engines were always going to be too heavy, and internal combustion engines with sufficient powers would not appear for almost 50 years. Other rotorcraft experiments continued, but advances in glider design captured the imagination of most designers, who believed fixed-wing aircraft were the future.

LEFT: **A model of the "Aerial Carriage" designed by Sir George Cayley in 1843. The steam-powered machine was fitted with two lifting rotors, two rotating wings and two pusher-type propellers. The machine was never built.**

LEFT: **Paul Cornu, seated in his helicopter in 1907. The two rotors rotated in opposite directions to cancel torque. The machine was the first to have risen from the ground in free flight, using rotor blades instead of wings.**

Early pioneers

Many inventors and engineers from around the world contributed to the research and development that led to the design of the first successful helicopter.

Enrico Forlanini (1848–1930) was an Italian engineer, inventor and aeronautical pioneer who carried out considerable research into helicopters. In 1877 he developed a helicopter design in which superheated, high-pressure steam drove a pair of two-bladed, contra-rotating rotors. The steam was created as the helicopter was on its "stand" on the ground and pumped into a metal sphere on the aircraft. As the high-pressure steam was released, it drove the rotors. A scale model was demonstrated in a park in Milan, where it rose to a height of 13m/43ft from a vertical take-off, and hovered for 20 seconds.

In 1906, brothers Louis and Jacques Bréguet began helicopter experiments and meticulously tested airfoil shapes for rotors under the guidance of Professor Charles Richet. In 1907, they built the Bréguet-Richet Gyroplane No.1, which was one of the first recorded mechanical devices to actually hover. Their gyroplane had a small engine that provided just enough power to enable it to fly for around one minute in the late summer of 1907, in what is generally accepted as the world's first vertical flight. As there was no means of control or stability, it required four men to steady it while it hovered 60cm/2ft off the ground.

A year later, again in France, bicycle maker and engineer Paul Cornu (1881–1944), was the first person to design and build a helicopter to achieve flight while carrying a passenger. Cornu's twin-rotor craft flew for some 20 seconds on November 13, 1907, rising to 30cm/12in in the air. A 24hp engine powered the helicopter, which had counter-rotating rotors. It is worth noting that both Cornu and Forlanini had both realized that without a tail rotor, a helicopter design would just spin on the rotor axis unless contra-rotation was employed. One can only imagine the countless heroic failures that led to this realization. Cornu, however, realizing that his design had no means of control in terms of forward or sideways motion, effectively reached his own technical dead end and abandoned this design after just a few flights.

Raul Pateras Pescara (1890–1966) was an Argentine lawyer and inventor who worked on seaplanes, engines compressors and helicopters. In 1919, Pescara built several true coaxial helicopter designs, which he described in the associated patent documents as a "rational helicopter". His designs were indeed

ABOVE: **Raul Pateras Pescara refined a coaxial rotor system that eliminated torque. This is the Pescara No. 3 machine.**

ABOVE: **The design by Corrandino D'Ascanio had two counter-rotating coaxial rotors, each with a trailing elevator to vary the rotor blade angle of attack.**

LEFT: **Louis Brennan conducted propeller-driven rotor experiments from 1919 to 1926.**
BELOW: **The Brequet-Dorand Gyroplane was flown for the first time in 1934. The machine had controls for collective and cyclic pitch.**

pure helicopters, at least in theory, and from 1919 to 1923, no doubt benefitting from his own legal knowledge, he submitted around 40 related patents in several countries. Meanwhile, he had not neglected the technical developments at all, and refined the practical applications of his designs. His first machine weighed around 600kg/1,323lb without a pilot, and power was provided by one 45hp Hispano engine. Each of the two coaxial rotors was made up of 12 propeller blades (i.e. 24 in total), but the small engine simply could not lift the weight of the machine. A modified design with a much more powerful Le Rhône rotary engine did manage to just lift off the ground in May 1921.

Pescara's most successful design, the No.3 built in 1923, was, by January 1924, making flights of around 10 minutes' duration. The same coaxial rotor system was then fitted with a Hispano-Suiza engine providing the power. On April 18, 1924, Pescara achieved a new world record, with a flight of 736m/2,419ft covered in 4 minutes, 11 seconds (a speed of approximately 13kph/8mph) at a height of 1.8m/6ft. Most significantly, Pescara achieved forward motion by altering the pitch of the rotor blades in flight by warping, and the rotor head could be tilted to give the blades a degree of forward thrust. Pescara had successfully demonstrated the principles of cyclic and collective pitch control. Impressively, he appreciated how to make use of autorotation when his engine failed. Pescara never capitalized on his achievements in the helicopter world, and shifted his attention to the motor industry. His patents and associated royalties may, however, have been reward enough.

Etienne Oemichen was another French pioneer who began his experiments in 1920 by cleverly floating a balloon above a twin-rotor helicopter to provide extra lift for the machine. A subsequent Oemichen design had four lifting airscrews and five auxiliary propellers, and he flew this on April 14, 1924, powered by a 134kw/180hp Le Rhône engine, establishing a 360m/1181ft distance record – the first officially recognized by the Federation Aeronautique Internationale (FAI). Then, a few weeks later on May 4, Oemichen became the first person to fly a helicopter at least 1km/0.6miles, to set a closed-circuit record of 1.7km/5,550ft on a flight that reached a height of 15m/50ft and lasted 7 minutes, 40 seconds.

Corrandino D'Ascanio (1891–1981) was fascinated by aviation from an early age and after qualifying as a mechanical engineer, he joined the Italian Army where he was soon involved in military aviation. After World War I he began to consider the challenges of control in helicopters, and filed a number of related patents. In 1925, he co-founded a company which, in 1930, produced the large D'AT3 that had two double-bladed, counter-rotating rotors. Control was achieved by using auxiliary wings or servotabs on the rotor blade trailing edges. Additional control of pitch, roll and yaw came from three small propellers mounted on the airframe. The D'AT3 set international speed and altitude records. The Depression halted helicopter research and D'Ascanio went to work for Piaggio, developing high-speed adjustable-pitch propellers. During World War II, having been made a General in the Regia Aeronautica (Italian Air Force), he restarted helicopter development in 1942. Post-war Italy was forbidden the research or development of military and aviation technology, so D'Ascanio became involved in the development of a cheap, easy-to-manufacture motor scooter, and designed the iconic Vespa scooter of which many millions have been built since 1946.

LEFT: **Juan de la Cierva designed and built his first autogyro, the C.1, in 1920. He used the fuselage from a French-built Deperdussin monoplane, and fitted his own rotor system. The machine did not fly, but during ground trials it did prove the concept of autorotation.**

Juan de la Cierva

Juan de la Cierva was born in Murcia, Spain, in 1895, and he is acknowledged as the inventor of the autogyro. He began designing and building aircraft from the age of 17, when he completed rebuilding a Sommer biplane which he re-engined and improved, naming it the BCD-1 El Cangrejo (crab). The aircraft is considered to be the first to be built in Spain. His final fixed-wing design was the C-3 built for a 1918 Spanish military competition for new aircraft. The large tri-motor biplane was completed in May 1919, but crashed in testing. The pilot had flown the aircraft too slowly, and it had stalled before crashing. A disappointed de la Cierva was inspired to consider alternative ways of flying at low speeds, and experiments with model helicopter designs led him to develop what he called, and trademarked, the "Autogiro". In de la Cierva's autogyro designs, the rotor was drawn through the air by a conventional propeller driven by an engine, while the rotor generated lift to sustain level flight, climb and descend.

In 1920, de la Cierva's journey to creating a viable autogyro design began with the Cierva C.1, which was a Deperdussin fixed-wing aircraft mounted with two contra-rotating rotors that provided lift and, by counter-rotating on a single drive shaft, helped eliminate torque that would affect stability. In testing, the C.1 simply would not fly, however it proved the principle of autorotation as it taxied.

BELOW: **The experimental C.8 autogyro was flown in the 1928 King's Cup Air Race before being used to make demonstration flights around continental Europe. The machine was flown to Paris from London in September 1928, and is now preserved at the Musée de l'Air et de l'Espace, Le Bourget, France.**

LEFT: **The C.6 was Juan de la Cierva's first successful design. The machine utilized the fuselage of a British-built Avro 504 biplane, and was powered by a French-built Clerget 9B rotary piston engine. The C.6C and C.6D were respectively manufactured by A.V. Roe & Company (Avro) as the Type 574 and Type 575.**

His next design was the C.2, which had only one rotor with five blades. This was put on hold as funds ran out, so de la Cierva designed the Cierva C.3, which was completed in June 1921. It had a three-blade rotor as well as a rudder and elevator for yaw and pitch control, while lateral control was dictated by what came to be known as collective pitch. This meant the angle of attack of the rotor blades would be changed at the same time. The design was shown to be too complex, and the C.3 only made a few short flights. He then revisited the C.2. which was finally completed in 1922. While lateral control was improved, sustained flight still eluded the autogyro pioneer.

These early experiments had fallen foul of asymmetric lift, which caused the aircraft to tilt during take-off. Lift between the advancing and retreating rigid rotor blades was not equal as they rotated, and de la Cierva's brilliant solution was the development of the flapping hinge. This allowed the rotor to rise and fall depending on the direction in which the blades were moving. Blades moving with the aircraft in forward motion rose because of the higher lift, but this also served to decrease their angle of attack. The blades travelling in the opposite direction to the autogyro would fall because of the lower lift, which increased their angle of attack. The rising and falling action, known as flapping, and the resulting increase and decrease of the angle of attack balanced the lift created on each side of the aircraft. Hinged blades also eliminated the gyroscopic effect caused by the rigid blades.

The Cierva C.4 incorporated hinged rotors, ailerons mounted on outriggers to the side of the aircraft for lateral control, while pitch and yaw control came from rudder and elevators. On January 17, 1923, the C.4 flew successfully, and performed the first controlled flight of an autogyro. Just three days later, an inflight engine failure demonstrated just how safe the autogyro could be, compared to fixed-wing aircraft. During the flight, the C.4 went into a steep nose-up attitude after the engine failed when the aircraft was just 11m/35ft off the ground. While a fixed-wing aircraft would have stalled and probably crashed, the C.4 beautifully demonstrated the principle of autorotation, and slowly and safely descended to the ground.

The Cierva C.6 had a four-blade rotor with flapping hinges, and used an Avro 504K fuselage and fixed-wing aircraft controls for pitch, roll and yaw. Having conducted his groundbreaking work in Spain, and following a successful demonstration tour with the C.6 in 1926, de la Cierva relocated to Britain where, in partnership with the Scottish industrialist James G. Weir, he established the Cierva Autogiro Company. He then focused on designing and producing the rotor systems, while the airframes were built by other aircraft constructors.

His autogyro designs continuously improved, and when fitted with more powerful engines, achieved higher performance. He had to overcome innumerable technical challenges along the way – as the autogyro pioneer, he was effectively writing the rule book as he went. In the Cierva C.19 the problem of getting the rotors up to speed was finally solved with a direct drive from the engine to the rotor, which was then disconnected prior to the take-off run.

As his autogyros achieved success, they inspired other engineers, who were then able to bring their expertise to benefit de la Cierva. His autogyros, as well as serving with the Royal Air Force in World War II, were built under licence in many countries, including France, Germany, Japan, the Soviet Union and the USA. Juan de la Cierva died in the crash of a Douglas DC-2 shortly after take-off from Croydon Airport on December 9, 1936, but his technical legacy was considerable.

ABOVE: **The experimental C.29 had an enclosed cabin that seated five, and was the heaviest and largest autogyro produced by Cierva at that time.**

LEFT: **Test pilot Frank Courtney (to the right of the aircraft wearing goggles and cap), talking with officials and reporters at Farnborough Aerodrome before taking the Cierva's sixth autogyro design for a flight on October 19, 1925.**

Autogyros of the world

Juan de la Cierva's work influenced autogyro designs across the world. The first autogyro made in the Soviet Union was the KASKR-1 "Red Engineer" based on the Cierva C.8. and, like the C.8, featured an Avro 504 fuselage. Built by Kamov and Skrzhinskii (hence the KA plus SKR in the designation), the machine was first tested in September 1929, and proved to have control and power issues. It was rebuilt as the KASKR-2 with a more powerful engine, and was flown successfully in 1931, leading to later Soviet rotorcraft development.

In the USA, the Kellett Autogiro Corporation was set up in 1929 with a licence from Cierva. Early models were basically the same as those produced in the UK. The KD-1, the first non-experimental rotary-wing aircraft in US Army service, was very similar to the Cierva C.30, but with a US-built engine.

The requirements of the US military called for Kellett to revise the design, so the YO-60 with an enclosed cockpit was very much an original design. The KD-1B (civil version of the YO-60) was used by Eastern Airlines to launch the first US rotary-wing airmail service in July 1939.

In Germany during the mid-1930s, Flettner was experimenting with both helicopters and autogyros. The Fl 184 autogyro, designed for the German Navy, was used for anti-submarine and reconnaissance missions, but the prototype caught fire in flight, and the company then resolved to focus on helicopter development. Elsewhere in Germany, the Focke-Achgelis company, having built up their rotorcraft knowledge as licensed builders of Cierva autogyros, were applying this experience to the building of experimental helicopters.

The Kayaba Ka-1, built in Japan, is worthy of note because it is believed to have been the first military rotorcraft to be armed and used operationally. In common with many military powers, the Imperial Japanese Army could see the value of the autogyro as army co-operation/artillery-spotting aircraft. In 1939, a Kellett KD-1A two-seat autogyro was imported from the USA for evaluation. Soon after its arrival in Japan, the autogyro was written off during trials, and what was left of it was handed over to Kayaba, who were conducting autogyro research. Employing some reverse engineering techniques, the US-built machine was studied closely, leading to the production of the Ka-1 two-seat observation autogyro. The Japanese-built machine was powered by a 240hp Argus engine (225hp Jacobs engine in the US-built machine) driving a two-blade propeller,

LEFT: **In 1935, the Kellett YG-1 (KD-1A) was flown on evaluation trials for the US Army at Langley Field, Virginia. A total of seven, fitted with an enclosed cockpit and powered by a Jacobs R-755 radial piston engine, were built for the US Army, and designated XO-60.**

LEFT: **The Pitcairn PAA-1 Autogyro was flown at Langley Research Center as part of the National Advisory Committee for Aeronautics (NACA) experimental rotor blade research programme.**

and had a three-blade rotor. In mid-1941, after successful testing, the Ka-1 was produced as an artillery co-operation aircraft. Once Japan was at war, the autogyro's performance also interested the Imperial Japanese Navy, who were suffering increasing shipping losses from Allied submarine actions. To counter this, the Ka-1 was modified for anti-submarine duties, operating from converted merchant ships and land bases. Changes included the deletion of one of the crew positions to allow space for carrying a small depth charge.

Developments by the US company Bensen attracted the interest of the post-war US military. Led by Russian immigrant Dr Igor Bensen who, having studied captured wartime German rotorcraft, produced the Bensen B-7 "homebuild" rotorkite. In late 1955, Bensen tested the powered B-7M, a true autogyro. Refinements led to the B-8M that was produced in its thousands for civilian use. As the X-25, it was considered by the US Air Force, which was losing valuable aircrew over Vietnam and wanted to examine integrating an autogyro or rotorkite into ejection seats. The Discretionary Descent Vehicle (DDV) programme would allow aircrew to deploy the craft as they descended, and thereby have more control over where they would land. No full-scale tests ever took place, and the end of the Vietnam War also ended the programme's proposed imaginative use of the autogyro principle.

BELOW: **The AV Roe-built Cierva C30A was produced in 1936, and was powered by an Armstrong Siddeley Civet radial piston engine. This aircraft entered Royal Air Force service as DR624.**

LEFT: **The rotor on the W.2 was linked to the engine by an auxiliary driveshaft. This allowed the rotor to be spun to give an improved take-off performance. The sole surviving W.2 is on display at the National Museum of Flight, East Fortune, Scotland.**

James George Weir

Air Commodore James George Weir (1887–1973) was a pioneering Scottish aviator and industrialist who financed Juan de la Cierva's development of the autogyro, and helped establish the Cierva Autogiro Company Limited.

Weir had served as a British Army officer from 1906, for part of that time in the Royal Artillery, and this experience would have given him a true appreciation of the value of using aircraft for battlefield observation and artillery spotting. On November 8, 1910, Weir was awarded Royal Aero Club Aviator's Certificate No. 24, having taken his test in a Blériot Monoplane at Hendon. This aviation experience led to his transfer to the embryonic Royal Flying Corps (RFC). He remained a serving officer until his retirement from the military in 1920.

In 1926, Weir helped to establish and then became chairman and managing director of Cierva. He came from a family of industrialists – his father and uncle had founded G & J Weir Ltd in 1871, a company which was a major manufacturer of specialized pumping equipment for the shipbuilding industry.

During World War I, the company manufactured munitions and built aircraft, including the Royal Aircraft Factory FE.2.

While the Cierva company, backed by Weir, marketed and refined Cierva's designs, a parallel family of autogyros was created by the Weir company, using Cierva patents and their considerable industrial capacity to produce the machines at their Cathcart, Glasgow, factory, as well as the engines, without the anxiety of development costs. G & J Weir designed and built four experimental autogyros from 1933–37, after which the company went on to produce the twin-rotor W.5 and W.6 helicopters before development was halted by the increasing industrial demands of World War II.

In 1932, Cyril Pullin joined Weir's aircraft department as chief designer, specifically to develop single-seat autogyros. Pullin had been a very successful motorcycle racer, winning the Isle of Man TT in 1914 while working for Douglas Motorcycles. In the late 1920s, he formed a motorcycle company to manufacture the Ascot-Pullin 500 and the Pullin-Groom light motorcycle. As an inventor, he had developed a number of helicopter engine patents in the 1920s, which were listed by the Douglas company.

The first Weir autogyro was the W.1 of 1933, powered by a Douglas 0-75 Dryad engine. The W (Weir) prefix was used on rotorcraft through to the Cierva W.14 Skeeter. The improved single-seat W.2, powered by a Weir 0-92 Flat Twin air-cooled engine, designed by Pullin and G. E. Walker, first flew in March 1934. The design had some stability problems, and the addition of tail surfaces led to the W.3, powered by a Weir Pixie four-cylinder in-line engine.

The W.3 included a Cierva "jump-start" design feature, in which the rotor was engaged to the Weir Pixie engine through a direct-drive shaft to generate lift without forward propulsion. Once airborne, drive from the engine was then

LEFT: **An improved tailplane assembly, with three fins, was fitted on the W.3 to overcome the stability problems that had been experienced on the W.2.**

LEFT: **The two-seat W.6 retained the outrigger configuration of the earlier Weir design. During testing, the helicopter attracted interest from the British military establishment. The pilot was Raymond Pullin, eldest son of the designer.**

changed to the tractor propeller for forward flight. This was the last autogyro manufactured by Weir. From 1938, the company concentrated on helicopters, which they believed had greater development potential, especially for carrying passengers.

The first helicopter design was the small, single-seat W.5 made in part from autogyro components. It had a plywood box-section outrigger on each side of the fuselage, with two-blade rotors on each hub. Powered by a Weir Four engine, cooled by a specially designed engine-driven blower unit, the W.5 was first flown at Dalrymple, Scotland, on June 7, 1938, with Raymond Pullin, the eldest son of Pullin, at the controls. This helicopter, the first in Britain to fly successfully, pioneered the now-standard helicopter safety feature of autorotation, in which the rotor can be disengaged from the engine in the event of power failure. The W.5 was controlled only by cyclic pitch, and there was no collective pitch; vertical control was achieved by varying rotor speed. The W.5 could be flown at up to 161kph/70mph. More than 100 flights were made between June 1938 and July 1939, at which point development was halted.

When Dr J. A. J. Bennett was chief technical officer of Cierva from 1936–39, he developed a rotary-wing design, the C.41. This was revived in April 1946 by the Fairey Aviation Company Limited, where Bennett became head of the rotary-wing aircraft division. The C.41 was developed as the Fairey Gyrodyne.

The W.6 design retained the outrigger configuration of the W.5, but was improved by using steel tubing to construct the fuselage and outriggers. A three-bladed rotor was mounted on each outrigger. The machine was powered by a de Havilland Gipsy VI engine. This pioneering two-seat helicopter first flew on October 27, 1939, and one of the early passengers was Sir (later Marshal of the Royal Air Force) Arthur Tedder, who at that time was Director General for Research in the Air Ministry. This was a clear indication of the level of official interest in this helicopter, and in the military applications of rotary-wing craft in general. As World War II escalated, the Weir company ceased helicopter development to increase production of more vital equipment that would make a more immediate contribution to the war effort.

By 1944, however, the company was working on the design of the Cierva W.9, a large single-rotor helicopter that used jet thrust to counter torque effect. Like the earlier W.5, the W.9 did not have collective pitch and used rotor speed to control vertical movement. When the prototype crashed during testing in 1946, further development of the project ceased.

The Weir-Cierva team next produced the large and impressive W.11 Air Horse, first flown in December 1948. Proposed as a crop-dusting platform, among other uses, the helicopter was unusual in having three rotors mounted on outriggers, driven by a single Rolls-Royce Merlin engine housed in the fuselage. The largest helicopter of the time, the first W.11 crashed during testing, and the project was scrapped. As a result, G & J Weir Ltd decided to withdraw all funding, and all helicopter development plans were passed to Saunders-Roe (Saro), including those for the W.14 Skeeter. This machine is usually thought of as a Saro design, but was in fact the last of the Weir-Cierva helicopters.

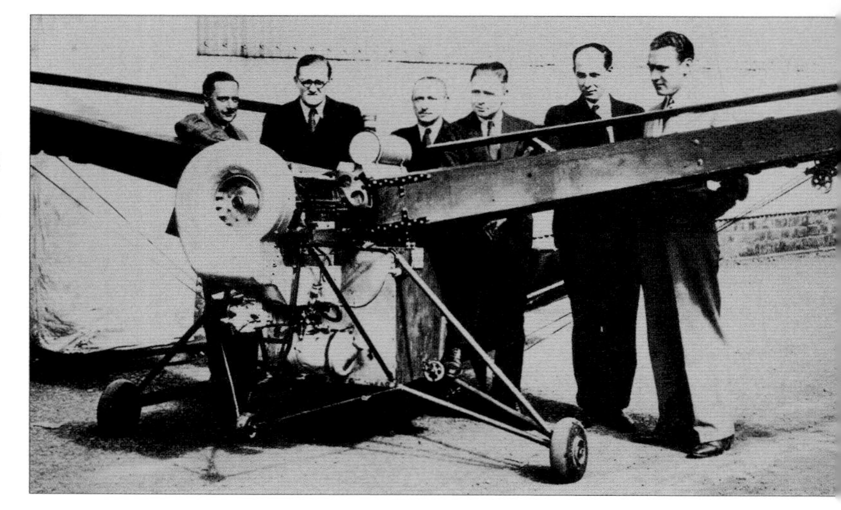

RIGHT: **First flown in June 1938, the W.5 pioneered the autorotation facility that is now standard on helicopters. This allows a safe descent in the event of engine failure. Note the plywood box-section outriggers housing the driveshafts, which transfer power from the engine to the rotors.**

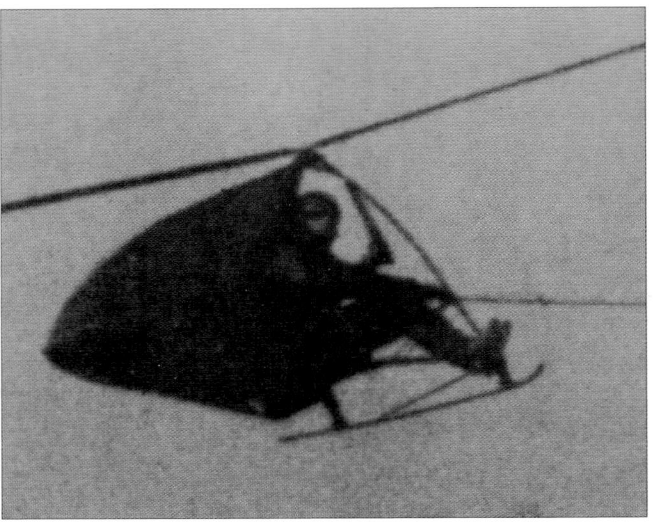

ABOVE AND RIGHT: **The experimental Rotachute was designed to enable paratroops to be deployed accurately in enemy territory. Towed into the air by a truck, the machine was tested extensively throughout 1942.**

Gyrogliders and rotorkites

A rotorkite, also known as a gyroglider, is a simple and unusual flying machine – essentially a towed, engineless autogyro in which the forward motion that turns the rotors is provided by a towing vehicle. Like an autogyro or helicopter, it relies on lift created by the rotors in order to fly. Considerable research into rotorkites and potential military applications was carried out during World War II, and one design, the Focke-Achgelis Fa 330, did see operational service with the German Navy.

German submarines sat low in the water and, as the crew could not see more than a few miles around even in perfect weather, were always at risk from fast-moving enemy warships. The Fa 330 Focke-Achgelis Bachstelze (wagtail) was developed to be towed by a winched cable in the air behind a U-boat, to extend the range of vision. By mid-1942, sea trials proved that the Fa 330 could work, but only the Type IX U-boat could tow the machine fast enough for flight, and only then in low-wind conditions.

The simple airframe consisted of two 6.35cm/2.5in diameter steel tubes forming an inverted "T". While one tube was the "fuselage" of the aircraft with the pilot's seat and instruments (altimeter, airspeed indicator and rotor tachometer), the other tube served as the rotor mast. The pilot simply moved the control column for direct pitch and roll control, and used foot pedals to move the large rudder to control yaw. The rotor blades consisted of a steel spar supporting plywood ribs skinned with fabric-covered plywood.

When not assembled, the Fa 330 was stored in two long watertight compartments built into the U-boat's conning tower. One tube contained the blades and tail, and the other the fuselage. In calm conditions, four crewmen could assemble the entire aircraft on the deck of a submarine in just three minutes.

As the U-boat moved ahead, the airflow would begin to spin the rotor, resulting in autorotation, the movement of relative wind through the rotor blades which caused them to turn with enough speed to generate lift and carry the machine aloft. To speed up take-off, a deckhand could pull hard on a rope wrapped around a drum on the rotor hub to spin the rotor.

The FA 330 was towed by a cable around 150m/492ft long, and "flew" 120m/394ft above the surface, where visibility was around 46km/29 miles compared to just under 10km/6 miles from the conning tower of a U-boat. Normal flight was 205rpm at a standard towing speed of 40kph/25mph, while a minimum

LEFT: **The Focke-Achgelis Fa 330 provided a simple solution to a serious problem facing German, and indeed all, wartime submarine commanders – the need to know what threats or targets may be over the horizon.**

speed of 27kph/17mph was required to maintain autorotation. The pilot talked to the submarine using an intercom system via a wire wrapped around the towing cable.

In the event of an attack that would cause the U-boat to dive, both the pilot and machine were expendable, although the pilot was equipped with a parachute. Allied air-cover was so good in the North Atlantic that only U-boats operating in the far southern parts of the Atlantic and the Indian Ocean deployed the Fa 330. Use of the Fa 330 assisted U-boat U-177 to intercept and sink the Greek cargo ship *Eithalia Mari* on August 6, 1943.

Meanwhile, in Britain, the talented Raoul Hafner was investigating alternative means of accurately deploying paratroopers. He led the rotorcraft team of the Airborne Forces Experimental Establishment (AFEE) at Ringway Aerodrome, now Manchester International Airport.

In October 1940, work began on the Rotachute, which was a simple steel tube frame where the pilot sat, open to the airstream, a two-blade 5m/15ft rotor, a rubber-mounted skid and a tapered fairing behind the pilot that stabilized the machine in flight. The self-inflating rear fairing was made entirely of rubberized fabric, typifying the budget approach to the design, which could have been produced cheaply in vast numbers. The rotor hub was rubber-mounted, which dampened vibration, and two-axis control was by a single lever fixed to the hub.

Tests with ever-larger models proved the concept, and one of the launch options considered was to release a number in a stream from the top of a large aircraft. The Rotachute itself weighed just 23kg/50lb and could carry 109kg/240lb, which could have included a soldier, a machine-gun, ammunition and

ABOVE: **Simple to manufacture and easy to assemble, the Fa 330 was stored in containers on the outside of the U-boat conning tower.**

a parachute. The size of the rotor made the Rotachute the smallest man-carrying vehicle capable of controlled flight built at that time. The first full-size test flights took place in early 1942, and refinements and improvements led to the Rotachute III that, by the end of 1942, was reaching altitudes of up to 1,189m/3,900ft and flight durations of up to 40 minutes. The Rotachute was never put into service.

The same team went on to develop the Rotabuggy, essentially a Jeep-type military vehicle combined with an autogyro. A 12m/41ft diameter rotor was attached, along with a tail assembly for stability in flight. Trials began in November 1943, and the Rotabuggy was eventually to reach glide speeds of 72kph/45mph. Free flight trials after being towed aloft by a Whitley bomber saw the machine reach a speed of 113kph/70mph at an altitude of 122m/400ft for up to ten minutes. Despite progress, the development of large vehicle-carrying assault gliders rendered the Rotabuggy obsolete, and the project was abandoned. The war ended before the concept of a Rotatank could be developed. This very heavy machine with a rotor diameter of 46m/152ft would have required two aircraft in tandem to tow it, but the design never progressed beyond the prototype stage.

ABOVE LEFT AND LEFT: **The Rotachute team went on to develop the Rotabuggy, a Jeep-type vehicle fitted with a rotor. A Rotatank concept never saw service.**

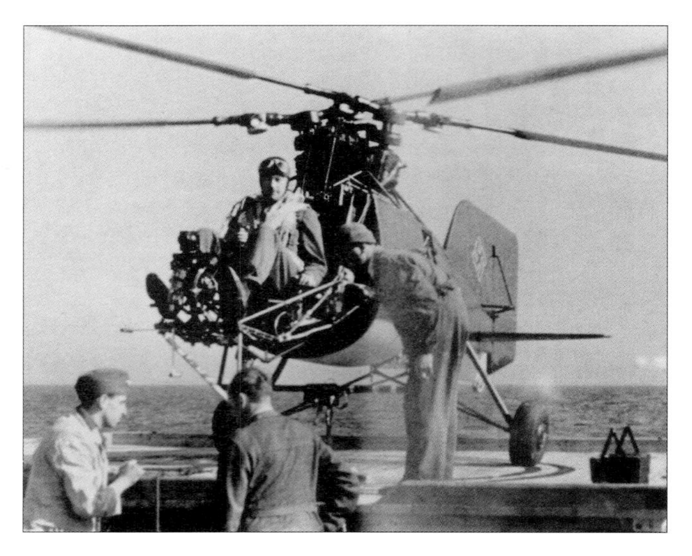

ABOVE: **The single engine of the Focke-Achgelis Fa 223 Drache (dragon) drove two three-blade rotors.** LEFT: **The Flettner Fl 282 Kolibri entered German Navy service, and was flown off the gun turret of a German warship.**

Helicopters in World War II

It is an often overlooked fact of aviation history that a small number of helicopter types did see active service, on both sides during World War II, and were far more than just experimental machines. The service of the Sikorsky R-4 is described elsewhere in this book, but two German designs are worthy of mention here.

Germany explored and developed a number of helicopter design beyond the ubiquitous Cierva-derived autogyros that many nations utilized in some form. Impressed by the potential of the Focke-Wulf Fw 61, the German Air Ministry instructed Henrich Focke to develop the Fw 61 design into one that could carry a 700kg/1,500lb payload. Having established the new company Focke-Achgelis with pilot Gerd Achgelis, they designed an enlarged Fw 61, initially designated Fa 226 Hornisse (hornet), that could carry up to six passengers. This design, the first-ever transport helicopter, was ordered by German national airline Lufthansa in 1938, before its first flight. In 1939, the type was then redesignated Fa 223, and entered military service as the Drache (dragon).

Although it retained the twin-rotor arrangement of the Fw 61, the Fa 223 had a fully enclosed cabin and freight/luggage bay, and was powered by a single BMW Bramo engine mounted in the centre of the tubular-steel airframe. The V1 prototype's first untethered flight took place in August 1940, and testing showed it to have the best performance of any helicopter to that date (it had a top speed of 182kph/113mph, a climb rate of 528m/1732ft per minute, and attained an altitude of 7,100m/23,300ft). The Fa 223 Drache was, however, a long way from entering front-line service, as the rulebook for operating helicopters effectively had yet to be written. Although numerous variants were proposed, just one multi-purpose version was developed. The pilot and observer sat side-by-side, and two steel tube outriggers extended from the fuselage side to carry the two non-overlapping three-bladed rotors.

BELOW: **The Sikorsky R-4 was the world's first large-scale, mass-produced helicopter, and the first helicopter to enter service with the US military, as well as the Royal Air Force and Royal Navy.**

ABOVE: **The Doblhoff WNF 342 was developed for the German Navy. The type was powered by a petrol engine which drove a small propeller and an air compressor to supply air to the pulse jets on the tips of the rotor.**

Production was consistently disrupted by Allied bombing, but the type's capability was proven by demonstration flights, including the lifting of a complete Fiesler Storch (stork) aircraft and operations over 1,600m/5,200ft above sea level and over great distances. In January 1945, three Fa 223s were assigned to the Luftwaffe's only wartime operational helicopter unit, Transportstaffel (Transport Squadron) 40 (TS/40) in Bavaria – two of these pioneering machines were seized by the Allies at the end of the war, and these or derivative machines were flown post-war by the USA, Britain, France and Czechoslovakia.

The Flettner Fl 282 Kolibri (hummingbird) was developed by German aeronautical scientist and helicopter pioneer Anton Flettner. Having built his first helicopter in 1930, Flettner produced the Fl 265 that first flew in 1939, and from this developed the Fl 282, designed from the outset for military use. Intended to carry a pilot and an observer, the design was judged to have so much potential for naval use that no fewer than 30 prototypes and 15 pre-production machines were ordered simultaneously to accelerate development and production.

The pilot sat in front of the rotors in a typically open cockpit while the observer sat in a single compartment aft of the rotors, facing backwards. Small, fast and agile, Luftwaffe fighter pilots found it hard to keep the small helicopter in their gunsights in mock attacks. The Fl 282 could land on a ship, even in heavy seas. Mass production was ordered, but Allied bombing of the factories meant that only the prototypes were produced. Nevertheless, 24 of these aircraft entered service with the German Navy in 1943 for escort service, flying off the gun turret of a ship to spot submarines, and to perform resupply missions in even the worst weather conditions. The Fl 282 was designed so that the rotor blades and landing gear could be removed and the helicopter could be stowed on a U-boat, although it is not known if this happened. This pioneering military helicopter served in the Baltic, North Aegean and the Mediterranean. Only three of these helicopters survived the war, as the rest of the machines were destroyed to prevent capture by the Allies. Two of the surviving machines went to the USA and Britain, while the third went to the Soviet Union. Anton Flettner moved to the USA after the war, and became chief designer at the Kaman Aircraft Corporation.

ABOVE: **A Flettner Fl 282 V – one of 30 prototype machines built.**
BELOW: **This production version of the Fl 282 was fitted with a rear cockpit to carry an observer.**

LEFT: **Igor Sikorsky, on the right of this photo in the light shirt, led his development team by example.** ABOVE: **Sikorsky at the controls of the Vought-Sikorsky VS-300 for its first tethered flight on September 14, 1939.**

Igor Sikorsky

Igor Ivanovich Sikorsky (1889–1922) was one of aviation's true pioneers, whose talents led to extraordinary developments in both fixed-wing and rotary aviation. He was born in Kiev on May 25, 1889, to a mother and father who were respectively a physician and psychology professor. He was absorbed by science and aviation from an early age, and his parents encouraged his interests. As a boy he built and flew model aircraft and helicopters, while hearing of the trailblazing work of people such as the Wright brothers. He attended the Imperial Russian Naval Academy, and went on to study engineering in Paris and then mechanical engineering in Kiev. His later studies were, however, curtailed by his passion for all things aviation, which took him back to Paris to learn more of the embryonic science of aeronautics.

Back in Kiev, he built his first helicopters, but the two machines failed to lift their own weight. Frustrated by the lack of available engine power and the limitations of his knowledge of rotary flight at that time, Sikorsky then turned to fixed-wing aircraft. Success came swiftly, and having been appointed head of the aviation subsidiary of the Russian Baltic Railroad Car Works, he designed the world's first four-engine aircraft – Russky Vityaz. Sikorsky's ambitious design included an enclosed cabin, lavatory, upholstered chairs and an exterior walkway around the nose, where braver passengers could take an inflight stroll. An even-larger aircraft followed in 1913, the Ilya Muromets, of which military versions saw action in World War I – it was the first strategic bomber. Sikorsky moved to France when the Bolshevik Revolution gripped Russia, and was soon at work on a bomber aircraft. This was never built as the war ended so, eager to find an outlet for his skills and passion, Sikorsky moved to the USA in 1919.

He struggled to find a position and then, in 1923, with financial backing from fellow immigrants who knew of his reputation from Russia, the Sikorsky Aircraft Corporation was born. The company produced a series of pioneering, successful, record-breaking flying boats that were used on routes around the world. In 1929, the Sikorsky Company became a subsidiary of United Aircraft & Transport Corporation, the predecessor of United Technologies. Sikorsky, now firmly established in the industry, returned to the challenge of helicopters. Since his early experiments almost three decades earlier, rotary craft understanding had advanced considerably due to experimental work in many other countries around the world. Designers now had a good understanding of the basic principles, including the workings of transmission and reduction gear boxes, and the collective and cyclic pitch required

RIGHT: **The Sikorsky XR-4 hovering at the Stratford, Connecticut plant during its first flight. Note the original horizontal-mounted tail rotor and open space frame-type structure for the aircaft.**

LEFT: **The Sikorsky S-52 was operated by the US Navy and the US Marine Corps as the HOSS-1. The HOSS-1G was operated by the US Coast Guard.**

to control vertical and horizontal flight respectively. Sikorsky appreciated the vital importance of these dynamic components, and had the genius to build on these findings and developments. Over the years he had continued to record and patent ideas for possible designs and, in 1938, United Aircraft backed his plans to build a new helicopter, the Vought-Sikorsky VS-300. The airframe was manufactured from welded steel tubing and was not covered. The cockpit was open and not fitted with instruments.

On September 14, 1939, Sikorsky, wearing his trademark Homburg hat, flew the tethered VS-300 a few feet off the ground, and gave the world the first practical single-rotor helicopter. It is also interesting to note that the design was developed purely with company money and had no government support. Breaking new ground in helicopter development was not easy, and the design was far from right first time – the vibrating, clattering machine was nicknamed "Igor's Nightmare" by some of the ground crew for good reason. The 8.5m/28ft three-blade main rotor and the first anti-torque rotor at the rear, for example, were not sufficient to allow the pilot complete control. However, after a long process of methodical refinements, modifications and experimentation with a number of configurations, Sikorsky settled on a single main rotor with cyclic pitch control and an anti-torque tail rotor. By 1941, the VS-300 had broken all helicopter records, and the design was to characterize Sikorsky helicopters henceforth and effectively set the template for the vast majority of all the world's helicopters. The success of the VS-300 led to military contracts, and in 1943 large-scale manufacture of the R-4 (derived from the VS-300) made it the world's first production helicopter. The company went on to produce the R-5, R-6 and the S-51 series, and were soon famous as a specialist helicopter manufacturer. Igor Sikorsky received countless academic, scientific and industry awards acknowledging his remarkable contribution to aviation. He continued to work long after others would have retired, and died at home on October 26, 1972, at the age of 83, after working a normal day at his office.

BELOW: **In 1941, Sikorsky fitted floats (also called pontoons) to the Sikorsky VS-300, making the first practical amphibious helicopter and opening up another area of area application.**

LEFT: **The slender greenhouse-type cockpit of the Sikorsky S-51 was a distinguishing feature of the machine. The type was the first Western helicopter to be widely exported.**

Charles H. Kaman

Charles Huron Kaman (1919–2011), in common with many aviation pioneers, was passionate about making flying models as a teenager, and he set national duration records for hand-launched model gliders. His dream of becoming a professional pilot was thwarted by deafness in one ear, so the young Charles decided to enter into aviation by another route. He graduated top of his aeronautical engineering class in 1940, and went to work for Hamilton Standard in the propeller performance unit. It was there that he met Igor Sikorsky, and was inspired by what he saw of his pioneering work with helicopters.

All designers face the challenge of the helicopter's fundamental stability and control problems. Kaman's simple arrangement in which twin rotors were intermeshed – this would generate increased lift and eliminate the need for innovative solution to the problem of torque. He developed an all-new concept of rotor control based on servo-flaps – small tabs added to the trailing edge of each rotor blade to give the pilot the ability to change the angle of attack of the blades and improve stability. He also proposed a unique goal was to make helicopters safer and easier to fly. Working at home in the evenings and at weekends, Kaman devised an a tail rotor, as torque would no longer be an issue.

By 1945, he was ready to strike out on his own, and with finance of just $2,000 from two friends, he founded the Kaman Aircraft (later Corporation) in his garage at his mother's home in West Hartford, Connecticut. From these modest beginnings, Kaman built a billion-dollar company, producing helicopters with an enviable reputation that went on to set numerous performance and altitude records.

ABOVE: **Charles H. Kaman with the test rig he built to run his first full-scale rotor assembly. A gifted engineer, Kaman patented his unique servo-flap rotor control system while he was still a young man.**

Kaman's first helicopter, the K-125, was an intermeshing, contra-rotating twin rotor, and first left the ground on January 15, 1947, during a short tethered flight. Within a few months, the K-125 had its first free flight, in which the test pilot took the helicopter straight up to an altitude of 15m/50ft and was soon executing figures-of-eight in the air at speeds of up to 97kmh/60mph. Subsequent developments led directly to the hugely successful HH-43 Huskie helicopter, which flew with the US Navy, US Marine Corps, US Air Force and a number of overseas air arms. In December 1951, a modified K-225 became the first helicopter to be powered by a gas-turbine engine.

Kaman's later SH-2 Seasprite multi-mission naval helicopter which, interestingly, did not have intermeshing rotors, flew more than 1,000,000 hours in service with the USN

RIGHT: **The experimental K-125 was Kaman's first helicopter, and utilized intermeshing rotors as well as the patented servo-flap rotor system. First flown in January 1947, it led to a series of production helicopters.**

LEFT: **The HOK-1 Huskie was operated by the USMC during the Vietnam War for liason, observation and rescue missions.** ABOVE: **Many versions of the Kaman HH-43 Huskie served with the US military.**

in anti-submarine warfare, anti-shipping and search and rescue. Kaman's designs proved very popular with the US military, due in no small part to the reputation that the helicopters had for performance and quality. In 1954, Kaman developed the first twin-turbine helicopter; in 1957, he designed and produced the first electrically powered helicopter drone. Kaman was always interested in business diversification and, among other successful ventures, he used his knowledge of music and plastics to produce the famous Ovation series of guitars, and formed the Kaman Music Corporation.

Charles Kaman received many awards and distinctions for his contributions to aviation, including the National Aeronautical Society Wright Brothers Memorial Trophy, the National Medal of Technology and the US Department of Defense (DoD) Distinguished Public Service Medal. He was inducted into the United States Hall of Honor at the National Museum of Naval Aviation, and is an Honorary Fellow of the Royal Aeronautical Society in Britain.

The HH-43 Huskie was flown on more rescue missions during the Vietnam War than all other helicopters combined, and Kaman was especially proud of the fact that his helicopters had been used to save an estimated 15,000 lives in the second half of the 20th century.

LEFT: **In 1992, Kaman introduced the K-1200 K-MAX, the first helicopter specifically designed for repetitive medium-lift operations. Two K-MAX aircraft have been modified for unmanned operations to conduct military resupply or humanitarian missions in high-threat environments.**

Frank N. Piasecki

Frank N. Piasecki, pilot, aeronautical and mechanical engineer, and pioneer in the development of transport helicopters and vertical lift aircraft, founded the hugely influential Piasecki Aircraft Corporation and Piasecki Helicopter Corporation.

Piasecki was born in Philadelphia, USA, on October 24, 1919, the only son of Polish immigrants. After school, he went on to study mechanical engineering at the University of Pennsylvania, before earning his Bachelor of Science degree from the Guggenhiem School of Aeronautics of New York University in 1940. Prior to attending university, Piasecki had been employed by the Kellett Autogryo Company and the Aero Service Corporation, both based in Philadelphia. After he graduated, he became a designer at Platt-LePage Aircraft Corporation, and then later worked as aerodynamicist for the Edward G. Budd Manufacturing Company, Aircraft Division.

ABOVE: **The Piasecki HRP-1 Rescuer was named the "Flying Banana" by aircrews, due to the distinctive shape of the machine.**

During 1940, the 21-year-old Piasecki, along with other fledgling engineers from the University of Pennsylvania, founded the PV Engineering Forum, which they ran in addition to their day jobs. The Forum ultimately became what is known today as the Rotorcraft Division of the Boeing Company. Piasecki, always keen to be in the thick of testing and leading by example, flew the first PV Engineering Forum helicopter, the PV-2, on April 11, 1943. The PV-2 was only the second successful helicopter to fly in the United States, so Piasecki was a pioneer of the truest kind. This trail-blazing single-seat, single-rotor helicopter with anti-torque tail rotor used full cyclic rotor control with dynamically balanced blades, and is now

LEFT: **The PV-3 was procured as the HRP-1 Rescuer by the USN, USMC and the USCG. The airframe was fabric-covered, and the machine was powered by a Pratt & Whitney R-1340-AN-1 radial piston engine.**

LEFT AND BELOW: **The Piasecki PV-2 was an audacious attempt to create the rotorcraft equivalent to the family car – it even had a horn fitted to the left landing gear. The surviving PV-2 is preserved in a US museum.**

preserved by the National Air and Space Museum of the Smithsonian Institution in Washington, DC.

The technical achievement of this pioneering flight attracted the attention of the US Navy, who were aware of the potential that helicopters had for naval operations. The USN awarded Piasecki a contract for the construction of his ambitious design for a large tandem rotor helicopter capable of carrying heavy loads internally. In March 1945, just 13 months later, again with the intrepid Piasecki at the controls, the world's first successful tandem rotor helicopter, the XHRP-1, took to the air. This machine, nicknamed the "Flying Banana", was the first-ever helicopter designed for the US Navy, and it was the forerunner of the modern tandem rotor helicopters in service. There were a number of people experimenting with helicopter designs at this time, but few got it as right as Frank Piasecki who, in the XHRP-1, created a helicopter capable of carrying as many passengers as similarly powered fixed-wing aircraft, and three times the weight of any helicopter flying at the time.

Piasecki played a large part in proving the helicopter to be an equal of fixed-wing aircraft and not a poor relation. His innovative tandem rotor design is considered to be pivotal in transforming the helicopter from at worst a novelty, at best a small aerial observation platform, into an aircraft with extensive military and civilian applications. The tandem rotor configuration was Piasecki's signature, and led to the development of the CH-46, the US Marine Corps primary assault helicopter for four decades, and the mighty Chinook, which continues to play a front-line role for numerous air arms around the world.

In 1946, the PV Engineering Forum became the Piasecki Helicopter Corporation, with Piasecki as both President and Chairman of the Board of Directors. Piasecki initiated further design studies that led to remarkable production transport helicopters including the HRP Rescuer series, HUP Retriever series, H-21 and H-16.

The Piasecki Helicopter Corporation was sold to the Boeing Airplane Company and its name was changed to the Vertol Division, but in the late 1950s, Piasecki and his fellow original founders formed the Piasecki Aircraft Corporation (PiAC) to continue research work on new Vertical Take-Off and Landing (VTOL) aircraft. PiAC developed a series of unique

experimental aircraft, including the Sea-Bat, an omni-directional VTOL Remotely Piloted Vehicle (RPV). By having two pairs of tilting propellers with no cyclic control but with dual differential collective control, the Sea-Bat was omni-directional.

Piasecki was the first man to qualify as a helicopter pilot with the US Civil Aeronautics Administration prior to receiving a fixed-wing pilot's licence. He did most of the test flying on his early helicopter designs, and was awarded numerous patents and awards for his work as a helicopter pioneer.

Frank Piasecki, a true pioneer in the vertical aviation industry, and one of the original inventors of the helicopter, died on February 11, 2008, aged 88 years.

ABOVE: **Piasecki later developed the 16H-1 Pathfinder high-speed compound aircraft. It was part helicopter, part fixed-wing aircraft, and was powered by a tail-mounted ducted propeller ring tail.**

Soviet pioneers

The first Soviet-designed autogyro was the KASKR-1 built by the pioneering designers Nikolay Ilyich Kamov and Nikolai K. Skrzhinsky. It made its maiden flight on September 25, 1929, with I. V. Mikheyev as pilot and Kamov in the rear cockpit. While their first design resembled a Cierva C.8 autogyro, in 1930 the designers produced the KASKR-2, based on the KASKR-1. Members of the Air Force Scientific and Research Institute participated in the evaluation of this autogyro, which made around 80 test flights. Among the team working on the autogyro was the gifted Mikhail Mil, who was a student of the Novocherkassk Polytechnical Institute and a protégé of Kamov.

In 1931, Kamov began working with the dedicated autogyro design group as part of the Central Aero and Hydrodynamics Institute (TsAGI) for Soviet autogyro development. Within two years he was heading one of the brigades that made up the group, while Mil was heading up another.

Progress in the work on autogyros within TsAGI, in turn, had a noticeable influence on the speed of achievements in helicopter design within the Institute. In 1932, the first Soviet experimental helicopter, the TsAGI 1-EA, a single-rotor machine with a rigid main rotor, was developed and flown to altitudes in excess of 600m/1,969ft. This design was followed by a number of improved experimental helicopters.

While autogyro research aided helicopter development, the experience gained from helicopters in turn influenced autogyro design. The A-7 which first flew in 1937 owed much to the helicopter experience gained within TsAGI. At the beginning of 1940, aircraft factory No. 290 was established and tasked with autogyro production with Kamov as chief designer and director, while Mil was appointed as his deputy.

The industrial, economic and military pressure caused by the war highlighted the relative importance of the autogyro compared to bomber and ground-attack aircraft, however, so production was terminated. A massive amount of rotary aircraft design expertise had nevertheless already been accumulated, including main rotor blades with tubular steel spars, rotor head with articulated blade attachments, automatic flapping damping of the rotor blades and a system for rotor thrust control. Mil later stated that they were just one step away from creating a viable helicopter, and just needed to perfect a gearbox to transmit the drive from the engine to the main rotor. As World War II neared its end, it was the helicopter with its ability to hover, fly very slowly if required and land and take-off vertically that captured the imagination of the military of the victorious nations, including the Soviet Union – the autogyro was effectively forgotten, and Kamov and Mil concentrated on helicopter designs.

The Kamov OKB (design bureau) was established in 1945, with specific responsibility for developing a contra-rotating rotor system for helicopters. Kamov developed a way of successfully mounting two rotors on one mast. As each

TOP: **The Mil V-12/Mil-12 that first flew in July 1968 is the largest helicopter ever built. Only two prtotypes were built, and this is the second, preserved at the Russian Air Force Museum at Monino.** ABOVE: **The Kamov Ka-22 Vintokryl was a turboshaft-powered convertiplane that first appeared in public in July 1961. The engines mounted at the wingtips powered both rotors and propellers.**

LEFT: **The Kamov Ka-10 (NATO identifier Hat) was developed from the Ka-8 as a light observation helicopter. An improved version, the Ka-10M – fitted with two tailfins and the more powerful Ivchenko AI-4G piston engine – was built and first flown in 1950.**

rotor moved in opposite directions, this innovative solution neutralized the torque effect of the rotor shaft and removed the requirement for a stabilizing tail rotor.

Kamov first used the system on a production helicopter in the Ka-15, which entered Soviet Navy service in the early 1950s. The large Ka-18 was developed from the Ka-15, and saw service with Aeroflot. The coaxial rotor system was to be a signature feature on most of Kamov's designs. In 1947, Mil was appointed to head the Helicopter Laboratory at TsAGI. This later became the Mil Moscow Helicopter Plant, designing 15 types of helicopter in over 200 variants, which went on to be built in their thousands. It is estimated that every fourth helicopter in the world today is of Mil origin. Mil died in 1970, but the OKB continued using his name. In 1971, his Mil Mi-12 won the Sikorsky Prize as the world's most powerful helicopter.

Kamov continued to lead his business until his death in 1973. Proud of his company's achievements, he set up a company museum in 1971. Between them, Kamov and Mil exerted a profound and long-lasting influence on the world of aviation.

LEFT: **The Kamov Ka-226 (NATO identifier Hoodlum) is a small, twin-engined utility helicopter. In place of a conventional cabin, the type has an interchangeable mission cargo pod.**

ABOVE: **In the Bell 47, Arthur Young and his team created the world's first commercially viable helicopter and a design classic. Numerous manufacturers had previously rejected his designs.**

Arthur Young (Bell Aircraft Corporation)

Arthur Middleton Young (1905–95) was a brilliant and extraordinary character. He was an inventor, cosmologist, philosopher, astrologer and a helicopter pioneer. From an early age, Young grappled with philosophical matters such as the nature of reality, and decided that to develop the mental tools needed for this demanding intellectual pursuit he would first study mathematics and engineering. In his final year at Princeton University, Young devoted himself to philosophy and devising a comprehensive theory of the universe. Unable to develop his theory to his own satisfaction, a frustrated Young decided instead to set himself a goal in which solutions could be tested. Young visited the US Patent Office to consider areas of research that might interest him, including television and 3D film. However, it was the development of the helicopter that caught his attention – up to that point these machines had not been perfected, and Young was confident that research and testing could provide the solution. He devoted the next 19 years to working on helicopter design and development.

LEFT: **The Bell 47B was a record-breaking model which is now preserved by the Smithsonian Air and Space Museum. The machine was used in service for 57 years.**

After graduation in 1927, Young returned to the family estate in Radnor, Pennsylvania, where for the next 12 years he worked alone on developing and perfecting helicopter design. Young was a wealthy young man, so was not driven to cut corners or rush his development because of some financial imperative. At that time, helicopters were still considered to be something of a novelty, and no manufacturer would have supported all his years of research and experimentation. As a boy he had been a gifted and prolific modelmaker, and he tested his theories on a succession of model helicopters built in a barn converted into an aeronautical laboratory.

The models were made from hobby store supplies and were powered by elastic bands or small electric motors. By 1937, Young was using a 20hp outboard motor to power complex geared rotor assemblies. He experienced many technical failures but, as he said in a later interview, "the experience gained in calculating stress and building parts proved invaluable". He also appreciated the need for trial and error in trying to master helicopter stability: "this meant I had to have flights and have wrecks". Young wanted to solve as many technical issues as possible using models. This was how he had a major breakthrough and developed the stabilizer bar that kept his

LEFT: **The prototype of the XHSL-1 (Bell Model 61) first flew in March 1953, and was designed to meet an urgent US Navy requirement for an anti-submarine warfare helicopter. Development problems resulted in limited production and service.**

LEFT: **The Bell XV-3 (Bell Model 200) tilt-rotor aircraft was first flown in August 1955, and featured an engine mounted in the fuselage with drive shafts to transfer power to two-bladed rotor assemblies mounted on the wingtips.**

model machines and then full-size helicopters in stable flight. Young also developed a functioning remote control system that enabled him to fly the model around and out of his barn.

A number of manufacturers had turned down the chance to view Young's findings, until a friend's chance remark got him an appointment in 1941 with the Bell Aircraft Company in Buffalo, New York. Having presented his data and his models, Bell agreed to build two full-scale prototypes. Young and his small team were given a facility remote from the main Bell plant, where they could develop and test the revolutionary machine in secret.

The first prototype Bell 30 was built in just six months, and Arthur Young himself, though not a pilot, took the controls for the first tethered flights. On June 26, 1943, the tethering cable was removed, and pilot Floyd Carlson took the Model 30 on its maiden flight. Igor Sikorsky visited the plant to see Young's machine for himself, and when the Bell 30 gave public demonstrations, it caused a sensation, flying inside large buildings and demonstrating precision manoeuvres in front of stunned crowds. As the design was refined and World War II was drawing to an end, Bell Aircraft saw the helicopter as a way of keeping their peacetime order books full, so were keen to develop commercially viable helicopter designs.

Having perfected experimental machines, Young's team then focused on the design which became the Bell 47. On March 8, 1946, the type received Helicopter Type Certificate H-1 as the world's first commercial helicopter. The Type 47 had a 27-year manufacturing history, and over 5,000 military and commercial machines were built in 20 versions in the US and under licence overseas. It was considered to be such an iconic and influential design that an example was added to the permanent collection of New York's Museum of Modern Art in 1984.

Young felt that he had found the solution in developing a viable helicopter, so he left Bell in 1947, returned to philosophy, and founded the Institute for the Study of Consciousness in Berkeley.

LEFT: **The Hiller YROE-1 Rotorcycle was a single-seat ultralight foldable self-rescue and observation helicopter, designed in 1953 to a military requirement.**

Stanley Hiller

Stanley Hiller Jr. was born in San Francisco in 1924 and showed an early flair for technical innovation and business, having designed and produced a range of small, fast model racing cars powered by model aircraft engines. By the age of 17, his company Hiller Industries was producing 350 miniature cars each month, and turned over $100,000 a year. To improve the strength of his model cars, Hiller invented a die-casting machine that increased the strength of aluminium castings used in their construction. During his short academic career at the University of California at Berkeley, Hiller's technical innovation came to the attention of the US military, and Hiller Industries was soon producing aluminium parts for fighter aircraft. Ever the businessman, in his spare time Hiller designed cast aluminium kitchen utensils to keep his small factory busy when war-related orders were completed.

Hiller's father Stanley Sr. was also an engineer and inventor who had built and flown his own aircraft in 1910, so when Hiller was asked to reflect on his remarkable achievements from an early age, he said: "I was fortunate in my choice of a father." Hiller had been considering helicopter designs for some time. As a 15-year-old he had read about Igor Sikorsky's rotary wing experiments, and believed he had a solution to the torque-induced instability that plagued early helicopter designs. He believed that contra-rotating coaxial rotors were the answer and would eliminate the need for a tail rotor and associated complex drive mechanism. The teenager tested the theory by dropping a model coaxial helicopter from his father's ninth-storey office window, and proved his theory. His determination to develop the coaxial concept drew him away from university, and he focused on the production of a 45kg/100lb model which, when demonstrated, convinced the US military to commission Hiller to build a full-size coaxial helicopter, the XH-44 Hiller-Copter. Although he had never flown a helicopter before, or even seen one fly, on July 4, 1944, Hiller conducted the test flight of the XH-44, an all-metal single-seat helicopter powered by a 90hp petrol engine, coaxial rotors and no tail rotor. This first successful flight of a helicopter in the western US and the first-ever flight of a coaxial helicopter in the US made Hiller famous. He was hailed in the international press and became the youngest person to receive the Fawcett Aviation Award for major contributions to the advancement of aviation.

ABOVE: **The Hiller UH-5 that led to the Hiller 360 was initially unstable in trials until it was fitted with the patented Rotormatic Control System – two small paddles which acted as a control rotor.**

The Hiller business grew, and was associated with the giant Kaiser Company for a time. In 1945, he formed United Helicopters and focused on the post-war commercial helicopter market with the Hiller UH-4 Commuter helicopter, a two-seat personal helicopter with coaxial rotors.

Hiller then developed the Rotormatic Control System, which reverted to a single rotor that achieved stability with a greatly simplified tail-rotor configuration. This design was the basis of the influential Hiller 360, only the third helicopter certified by the US Civil Aeronautics Administration (CAA). By now operating under the name of Hiller Aircraft Company, it was the first company in the US to produce helicopters without military sponsorship, his backers sharing his view that the machine would revolutionize utility operations in, among

RIGHT: **On July 4, 1944, Hiller flew his XH-44 design and himself into aviation history – performing the very first flight of a coaxial helicopter in the USA.**

other markets, agriculture, crop spraying and rescue. Worldwide marketing led to orders from the French, who ordered the production version, the UH-12, for medical evacuation in the Indo-China war, and the type proved very capable under difficult jungle conditions. Hiller had been urging the US Army to consider the UH-12, and the outbreak of the Korean War finally led to large orders. Hiller Aircraft was soon delivering a helicopter a day for the Korean front line, and in US military service the type was designated the OH-23 Raven.

Stanley Hiller was a gifted leader who encouraged creativity in his team, so Hiller Aircraft had the reputation as the people to go to who could make often demanding operational requirements a reality. The OH-23 series became the first helicopter of any type to be approved for 1,000 hours of operation between major overhauls. As early as 1947, Hiller's creative group had experimented with rotor systems that tilted forward for higher speed horizontal flight, and the company was an early pioneer in pure jet lift concepts. By 1951, the company was flying the two-seat Hornet powered by ramjets (designed by the company) on the rotor blade tips.

Hiller Aircraft Corporation's imagination seemed to know no bounds, and other projects included the Hiller Flying Platform for the US Army, using ducted propellers to lift its operator, who only needed to lean in the direction he wanted to go. The Hiller XROE-1 Rotorcycle was an ultra-light one-man helicopter that could be parachuted behind enemy lines, assembled in nine minutes and take off to fly like a standard helicopter. In 1956, Hiller developed a high-speed vertical take-off and landing (VTOL) aircraft to transport troops and equipment into inaccessible combat locations. The tilt-wing X-18 was fitted with a wing that could pivot through 90 degrees so that the propellers would act like helicopter rotors. Although no orders resulted, the data generated were vital for later VTOL research projects.

ABOVE: **The Hiller YH-32 Hornet was powered by Hiller HRJ-2B ram jet units mounted on the tips of the rotor blades. A total of 14 were built for the US Army.**

In 1968, Stanley Hiller merged Hiller Aircraft into what became the Fairchild Hiller Corporation, and left aviation to set up a what was to become a hugely successful management consultancy that turned ailing companies around. He died aged 81 on April 20, 2006.

RIGHT: **The Fairchild Hiller FH-1100 light helicopter was designed for the US Army Light Observation Helicopter (LOH) competition. A Hughes design was chosen, but the type was successfully marketed by Hiller as the FH-1100 civilian helicopter.**

Sikorsky R-4 Hoverfly

The Vought-Sikorsky VS-316 was developed from the VS-300 and, under the US Army Air Forces (USAAF) system for Rotorcraft, was designated as the XR-4. The machine was first flown on January 13, 1942, and was delivered to the USAAF for evaluation on May 30, 1942. The R-4 was flown operationally during World War II over the jungles of the Burmese border region and areas of the South Pacific. The type entered service with the Royal Air Force as the R-4 Hoverfly. Many of these aircraft were to be passed to the Royal Navy to train Fleet Air Arm pilots to fly helicopters.

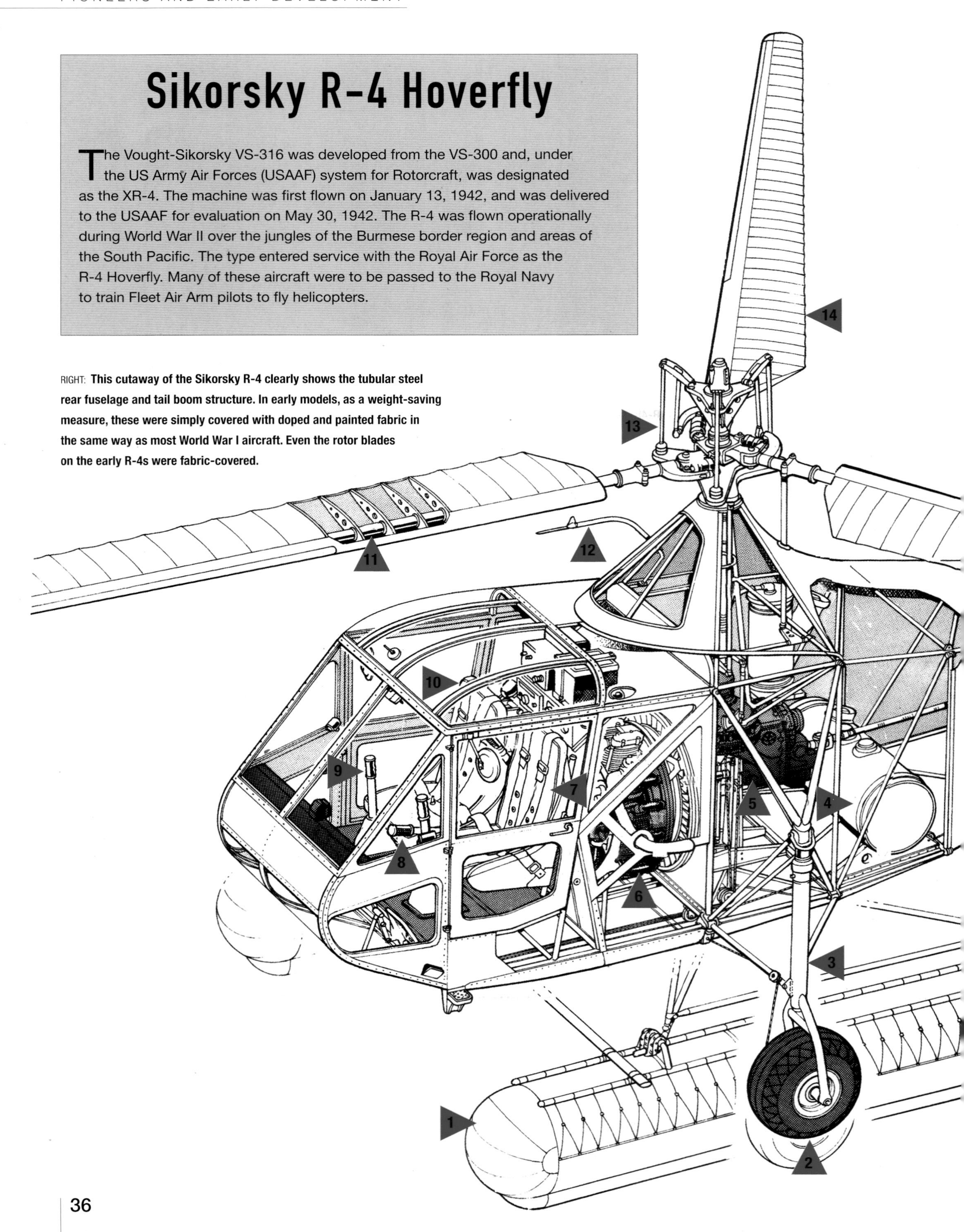

RIGHT: **This cutaway of the Sikorsky R-4 clearly shows the tubular steel rear fuselage and tail boom structure. In early models, as a weight-saving measure, these were simply covered with doped and painted fabric in the same way as most World War I aircraft. Even the rotor blades on the early R-4s were fabric-covered.**

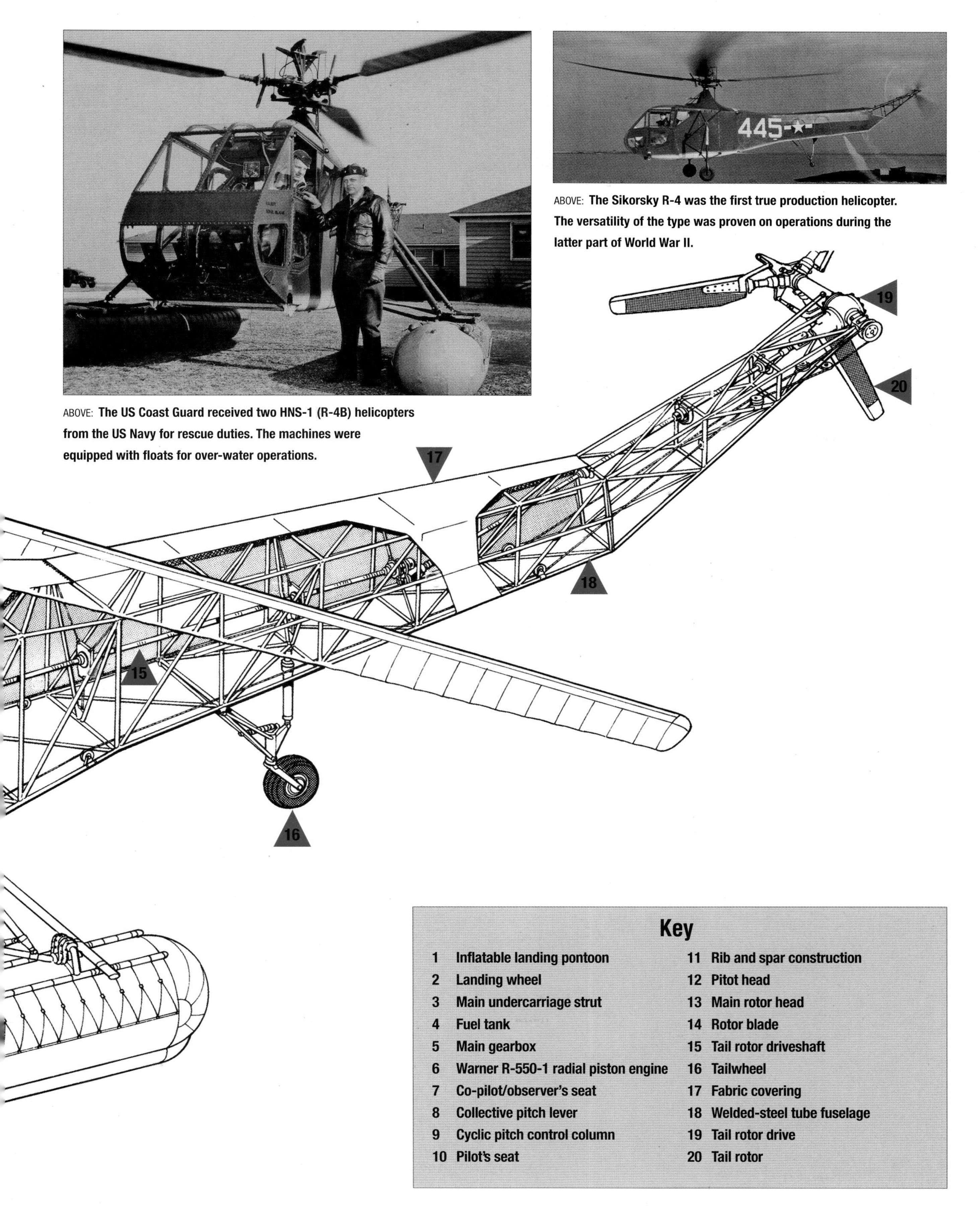

ABOVE: **The US Coast Guard received two HNS-1 (R-4B) helicopters from the US Navy for rescue duties. The machines were equipped with floats for over-water operations.**

ABOVE: **The Sikorsky R-4 was the first true production helicopter. The versatility of the type was proven on operations during the latter part of World War II.**

Key

1 Inflatable landing pontoon
2 Landing wheel
3 Main undercarriage strut
4 Fuel tank
5 Main gearbox
6 Warner R-550-1 radial piston engine
7 Co-pilot/observer's seat
8 Collective pitch lever
9 Cyclic pitch control column
10 Pilot's seat
11 Rib and spar construction
12 Pitot head
13 Main rotor head
14 Rotor blade
15 Tail rotor driveshaft
16 Tailwheel
17 Fabric covering
18 Welded-steel tube fuselage
19 Tail rotor drive
20 Tail rotor

ROYAL AIR FORCE
RESCUE
XZ594
RAF
RESCUE

The helicopter comes of age

From tentative steps during World War II, the Korean War saw widespread use of the helicopter for moving military forces and equipment to and from battles zones while also carrying out vital medevac missions. Soldiers could now be inserted precisely to map coordinates and extracted too, if required. This experience under fire shaped the early use of US helicopters in the Vietnam War, which came to be known appropriately as the Helicopter War due to the huge numbers of personnel and vast quantities of equipment that were moved throughout that theatre of war.

The conflict also saw the birth of the helicopter gunship and attack helicopters, the development of which influenced Cold War strategies. Helicopter gunships are now key assets in many air arms, and have been developed in parallel with missile-armed naval helicopters that can attack large ships and enemy submarines. While early naval helicopters were simply used to move supplies and undertake aircrew rescue missions, at the height of the Cold War many helicopter types were armed with nuclear weapons for attacks on enemy craft.

The use of heavy-lift and other large helicopters can prove to be game-changers in fast-moving battle combat scenarios where there is a need to move or insert troops rapidly. The helicopter's ability to operate without airstrips, from land or sea, means it has become one of the most important military assets available to the military commanders of today.

LEFT: **The Westland Sea King HAR.3 is a search and rescue version produced for the RAF. To provide more space in the fuselage, the cabin rear bulkhead was repositioned.**

LEFT: **The Sikorsky H-19 Chickasaw first flew in November 1949, and after rapid development entered operational service. The US Army's 6th Transportation Helicopter Company (THC) was equipped with 12 of the type, and arrived in Korea during March 1953. The machines were used for light transport and casualty evacuation.**

The Korean War

In early August 1950, when US Marine Corps forces were rushed into action to help defend the Pusan perimeter against a North Korean assault, their Brigade Commander was flown in a Marine Observation Squadron 6 (VMO-6) Sikorsky HO3S-1 helicopter to assess the terrain, find a location for a command post and visit unit commanders. Although British, US and German forces were all operating helicopters by the end of World War II, helicopters were still seen as something of a novelty, and critics focused on their limitations rather than their many unique strengths. The US military's use of helicopters in the Korean War (1950–53) saw the type prove beyond doubt to be a vital element in the inventory of any military force.

Four main helicopter types were used in the Korean War, some in a number of different versions serving with different arms of the US forces – the Sikorsky H-5 and H-19, the Bell H-13 and the Hiller H-23.

The Sikorsky H-5/HO3S was a four-seat utility helicopter used by the US Air Force, US Navy and US Marine Corps during the Korean War. First into action were HO3S helicopters, from the aircraft carriers USS *Valley Forge* (CV-45) and USS *Philippine Sea* (CV-47), positioned offshore to support the retreating UN troops. These machines operated as plane guards to recover aircrew from the sea in the event of a landing or take-off accident.

Helicopters were also assigned to USN cruisers and battleships, and began to be used for spotting and directing the fall of the ships' big gun shells on land; this dramatically boosted accuracy and effectiveness. The HO3S was also used for mine spotting. The USAF 3rd Air Rescue Squadron (ARS), based in Japan, possessed nine H-5s at the beginning of the war, and they were soon in use evacuating casualties. These helicopters were also being used to rescue downed aircrew from the sea, and in many cases snatching them back to safety under the very noses of the enemy. This hazardous form of mission resulted in the loss of a number of Sikorsky helicopters and their crews to enemy fire.

ABOVE: **The Sikorsky H-5, fitted with two external stretcher-carrying pods, enabled the US military to transport wounded troops back from the front line for treatment. The type also provided a new means of extracting personnel from behind enemy lines or from the sea before being captured.**

LEFT: **A Bell H-13 Sioux with personnel of a Mobile Army Surgical Hospital (MASH).**
BELOW: **A Sikorsky H-19 Chickasaw operated by the US Army is being used to deliver ammunition to a hill-top artillery position.**

The number of helicopters available in Korea far outstripped demand, and commanders were soon asking for the deployment of larger, more capable machines to move troops and supplies. The Sikorsky H-19 Chickasaw was an eight-seat utility helicopter used by the USAF, USN, US Army, and USMC during the war. Experimental YH-19s had arrived in Korea in March 1951, and within 24 hours were being used to evacuate casualties.

The USMC Helicopter Transport Squadron 161 (HMR-161), which had arrived in Korea on August 30, 1951, was equipped with 15 of the USMC version, the HRS-1. During the last year of the war, HMR-161 expanded the role of USMC helicopters, and used them for air assault, air transport to support an attacking force, and for tactical redeployments. Meanwhile, the US Army's 6th Transportation Helicopter Company (THC), equipped with 12 machines, arrived in Korea in March 1953, and these were used mainly for light transport and for casualty evacuation (CASEVAC). In this role, the H-19 could carry six stretchers and one medical attendant.

The two-seat Bell 47 was widely used by the US Army and USMC during the war. The USMC version, the HTL-4 (as the H-13), arrived in Korea as part of Observation Squadron 6 (VMO-6) and served in throughout the war. Four US Army H-13B Sioux from the 2nd Army Helicopter Detachment (AHD) arrived in Korea in December 1950, and became the first US Army helicopters to serve in Korea. Although the primary role was battlefield observation and reconnaissance, in the first month of operation alone, they were used to move over 500 casualties from the combat zone. This led to the development of the H-13C equipped to carry two stretchers externally. The 801st Medical Air Evacuation Squadron (MAES) evacuated more than 4,700 casualties in December 1950, and were awarded a Distinguished Unit Citation (DUC).

The three-seat Hiller H-23 Raven served in Korea in relatively small numbers, but also inevitably served in the CASEVAC role as well as a battlefield observation and surveillance platform.

While its tactical value is not in question, the importance of the CASEVAC helicopter in Korea cannot be overstated. The unforgiving and difficult terrain in question meant that the time taken and the nature of transporting the injured in vehicles would have undoubtedly resulted in the deaths of many wounded US troops. Instead, casualties could be quickly removed to a safe location such as a Mobile Army Surgical Hospital (MASH), or possibly a hospital ship moored nearby off the coast.

LEFT: **Some helicopters, such as the Sikorsky HH-3E Jolly Green Giant, have a watertight boat-like hull to enable the machine to land directly on the water.**

Air Sea Rescue (ASR)

Before helicopters were available for the role, if a person had to be rescued from the sea, this could have been achieved in a number of ways. First, boats could sail towards the person in need of rescue, provided they knew the location. Fixed-wing aircraft could search huge tracts of ocean looking for survivors and drop dinghies, emergency supplies and even rigid lifeboats, and then direct rescue boats. If the sea conditions were right, flying boats or amphibious aircraft could land in the sea close to the victims, pick them up and fly them to safety. The helicopter combines the best of all these approaches, and can perform in conditions where fixed-wing aircraft cannot, then hover over the casualty, lift them to safety and, if required, fly them to a ship or even to a hospital on land.

Air sea rescue (ASR), also known as search and rescue and sea air rescue (both of which are abbreviated to SAR), is the coordinated search for and then rescue of the survivors or victims of an emergency at sea. Military SAR was developed to save downed aircrew as well as the crews of surface vessels and submarines, but in peacetime these assets are frequently used to rescue civilians. As soon as working helicopters were introduced in the 1940s, their application in ASR was a priority for many naval commanders who were also quick to appreciate that helicopters could perform many other tasks when not involved in rescue operations. A Sikorsky S-51 was used to carry out the first successful helicopter rescue at sea on November 29, 1945, when test pilot Dimitry Viner broke off from a demonstration flight for the US Navy and lifted two seamen to safety from a ship in

ABOVE: **The PZL Anakonda is the SAR version of the PZL W-3 Falcon operated by the Polish Navy, and the first helicopter to be completely designed and manufactured in Poland.** RIGHT: **The Westland Sea King helicopter is the main type of rescue helicopter in service with the RAF and RN.**

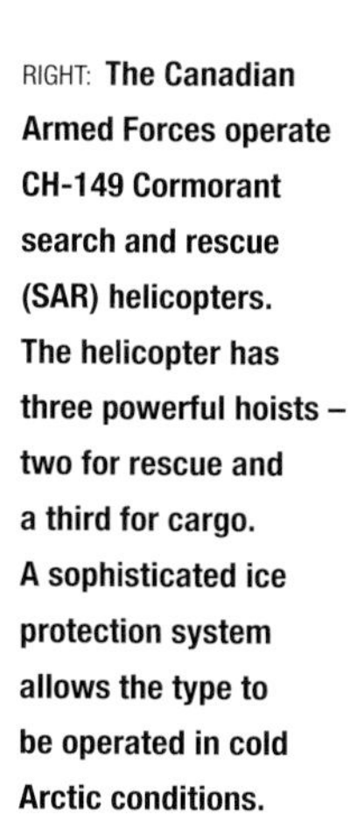

RIGHT: **The Canadian Armed Forces operate CH-149 Cormorant search and rescue (SAR) helicopters. The helicopter has three powerful hoists – two for rescue and a third for cargo. A sophisticated ice protection system allows the type to be operated in cold Arctic conditions.**

LEFT: **Pilots of the RAF and RN are currently trained at the Search and Rescue Training Unit (SARTU) at RAF Valley on Anglesey, using the Bell UH-1 designated HT-1 Griffin.**

distress not far from the Sikorsky airfield. Equally, helicopters are able to rescue people who may be trapped in inaccessible places – among rocks, for example.

The Korean War was the first conflict in which downed aircrew had a reasonable chance of being quickly found and rescued from the sea. In Britain, the Royal Air Force were keen to be able to rescue their aircrew, who may have parachuted into the seas around the UK. The world's first dedicated peacetime ASR unit was No. 275 Squadron, reformed for the task in 1953 at RAF Linton-on-Ouse. Equipped with Bristol Sycamore helicopters, No. 275 was tasked with providing ASR cover over the North Sea. The aircraft were painted bright yellow, which established a colour scheme that remains in use on many rescue helicopters around the world today. Other types joined the RAF's rescue services around the UK – 1956 saw the arrival of the Westland Whirlwind for the task. These were in turn replaced by the Westland Wessex HAR.2, which brought the added safety feature of multi-engine operations – especially reassuring when flying a search mission over large area of open sea. In August 1978, the Westland Sea King entered service, and after the Falklands War a number of machines from RAF Coltishall were deployed in support of RAF and Royal Navy fighter operations from Stanley Airfield, and were always on standby to rescue any downed British aircrew.

Although some rescue helicopters could land directly on water to carry out a rescue, due to a watertight hull, others, including some early Bell designs, were fitted with pontoons so that the helicopter could operate on both water and land. Most helicopters are equipped with a rescue winch and a crew member generally termed "rescue swimmer", who will jump into the water to help the survivor get into the hoisting harness.

Today, the Royal Air Force maintains a year-round, 24-hour search and rescue service operating from six locations covering the entire UK and large areas of coastal waters. For complex operations, the helicopters will operate in conjunction with a maritime patrol aircraft to coordinate a major rescue operation from the air.

LEFT: **The Fairey Jet Gyrodyne, a modification of the second prototype FB-1 Gyrodyne, was built to develop the pressure-jet rotor drive system.**

Gyrodynes

A gyrodyne is a type of rotorcraft that has a rotor system driven by an engine for take-off and landing, like a helicopter, but it also has one or two propellers mounted on the wing(s) for propulsion, while also having an anti-torque function. In forward flight, the spinning rotor provides lift and keeps the aircraft in the air.

Dr James Bennett developed the gyrodyne while he was Chief Engineer at the Cierva Autogiro Company, and patented the concept in 1939. Bennett's design study finally became a reality in 1945; when employed by Fairey Aviation, he updated and developed his concept and produced the Fairey Gyrodyne. It was a compact, aerodynamically refined rotorcraft weighing just over 2,000kg/4,410lb, and was powered by an Alvis Leonides radial piston engine that drove the rotor and a single starboard wing-tip mounted propeller as required. The rotor enabled the Gyrodyne to be hovered like a helicopter, while the propeller provided thrust for forward flight. A test flight on December 4, 1947, led to a period of intensive testing and evaluation, which included an attempt to set a new world helicopter speed record in a straight line. Just days before attempting another closed-circuit record in April 1949, a mechanical fault in the rotor caused the first prototype to crash, killing the crew.

The Gyrodyne had already secured an order from the British Army for use in Malaya, in preference to the Sikorsky S-51 Dragonfly and Bristol Sycamore, but the crash investigation delayed production, and the sales opportunity was lost. The heavily modified second prototype did not fly again until January 1954, and was by now called the Jet Gyrodyne, with the rotors driven by tipjets fed with air from compressors driven by the Alvis Leonides radial engine. Pusher-type propellers, now mounted at the tip of each stub wing, provided yaw control as well as thrust for forward flight. Fairey had not, however, built the Jet Gyrodyne as a new type, but used the machine to gather flight data in support of Fairey's Rotodyne project.

The Fairey Rotodyne was developed to satisfy the long-standing interest in a vertical take-off airliner that could operate from city centres and airports with equal ease. Fairey was confident that they had proved the concept with the Jet Gyrodyne, and built the Rotodyne which had short, fixed shoulder-mounted wings each with a Napier Eland turboprop engine for forward propulsion. The engines also produced compressed air which was fed to tipjets to be mixed with fuel and burned to drive the rotors for take-off, landing and

LEFT: **G-APJJ is the sole complete survivor of six Fairey Ultralights built, and is preserved at the Midland Air Museum in the UK. Air from the engine's compressor was bled to the rotor tips, where it was mixed with fuel and ignited.**

LEFT: **The Sikorsky X-2 was an experimental compound (gyrodyne) helicopter, which is thought to have been the fastest helicopter-type aircraft ever built. All performance data gained during flight testing were to be used for the Sikorsky S-92 Raider project.**

hovering. The rotors autorotated during flight cruise when all engine power was applied to the propellers. No anti-torque correction system was required, although propeller pitch was controlled by the rudder pedals for low-speed yaw control. Cockpit controls included cyclic and collective pitch levers, as found in a conventional helicopter.

The Fairey Rotodyne made its first flight on November 6, 1957, piloted by Chief Helicopter Test Pilot Sqdn Ldr W. Ron Gellatly and Assistant Chief Helicopter Test Pilot Lt Cdr John G. P. Morton. The first successful transition from vertical to horizontal flight and back was achieved on April 10, 1958. In testing, the Rotodyne also demonstrated that it could be hovered with one engine shut down, and also could be landed as an autogyro.

Tipjet drive and unloaded rotors produced a much better performance than that of a pure helicopter. The machine could be flown at 324kph/201mph, and pulled into steep climbing turns without demonstrating any adverse handling.

The proposed mode of operation for the Rotodyne was that it would take off vertically from a city centre location with all lift coming from the tipjet driven rotor, and then would increase forward airspeed with all power from the engines being transferred to the propellers while the rotor autorotated. In this mode, the wings would be taking as much as 50 per cent of the aircraft's weight. The Rotodyne would then cruise at 280kph/174mph to another city location, where the rotor tipjet system would be restarted for a vertical landing. When the Rotodyne landed and the rotor stopped moving, the blades drooped downwards from the hub. To avoid striking the tail fins on start-up, the fins were angled down to the horizontal, and would be raised up once the rotor was at running speed.

British European Airways (BEA), the RAF, New York Airways and the US Army were all interested in placing substantial orders for the Rotodyne, and a larger version was also being designed. However, increased costs, British aviation industry mergers, problems with engines, and finally a withdrawal of government funding doomed this very interesting project. Other companies have experimented with gyrodyne principles since then, and in recent years there has been a renewed interest in the concept.

RIGHT: **The Fairey Rotodyne was a compound gyroplane transport aircraft. High noise levels from the rotor tipjets spelt doom for this innovative project.**

LEFT: **HMS *Theseus* with Westland Whirlwind and Bristol Sycamore helicopters of the Joint Experimental Helicopter Unit (JEHU), which operated alongside Royal Navy helicopters. Note the French hospital ship in the background.**

The Suez Crisis

In 1956, the Suez Crisis erupted over Egyptian President Nasser's decision to nationalize the strategically important Suez Canal. The Anglo–French military response proved to be one of the final chapters for the British Empire. Britain's Prime Minister at the time was Anthony Eden, and when he died in 1997, *The Times* newspaper noted, "He was the last prime minister to believe Britain was a great power and the first to confront a crisis which proved she was not." The military actions are also remarkable because they included the first helicopter-borne assault landing.

LEFT: **The versatility of the helicopter proved invaluable during the crisis, being used to transport supplies to and from the invasion fleet.**

The Suez crisis began on October 29, 1956, when Israeli forces attacked Sinai. British and French combat aircraft carried out bombing and anti-shipping attacks over the following days as a prelude to an invasion of the Canal Zone. While some French and British troops were parachuted in and other British personnel came ashore on World War II-vintage landing craft, the Royal Marines of 45 Commando were taken into battle by a mixed fleet of Royal Navy, Army and Royal Air Force helicopters.

The Joint Experimental Helicopter Unit (JEHU), had been set up in 1955, and was the first joint British Army/RAF unit established since the end of World War II. Its role was to examine how best to transport troops by helicopter from RN ships. Having practised using a runway marked out as a carrier deck on dry land, the unit had first landed on a moving carrier in October 1955. The Suez crisis gave the unit an opportunity to put theory into practice, and it was positioned on board the light-fleet carrier HMS *Ocean*, complementing the helicopters of No. 845 Naval Air Squadron (NAS) on board HMS *Theseus*. The helicopters were also there, however, to serve in the rescue role. On November 3, a Westland Wyvern fighter-bomber was hit by Egyptian gunners during a ground-attack mission and the pilot had to eject, landing in the sea less than 5km/3 miles from an Egyptian shore battery. Fleet Air Arm (FAA) fighter aircraft mounted an air patrol overhead, while a Royal Navy helicopter was deployed to rescue the pilot from the sea.

At 16:00 hours on November 6, the helicopter crews on both carriers were given orders to begin the assault, and all were airborne within five minutes, as the first wave of the force of almost 500 commandos was carried to the landing zone at Port Said. Eight RN Whirlwind HAR.22s took part, together with a further six Whirlwind HAR.2s from the JEHU and six Bristol Sycamore HC.14s.

The helicopters returned immediately to the ships to pick up the next wave and, in around 90 minutes, 435 commandos together with some 2,337kg/5,152lb of equipment had been carried ashore. While the helicopters continued to deliver supplies, British and French casualties were flown back to the ships. One injured marine was back on board just 20 minutes after he had left, which illustrates the remarkable leap in casualty evacuation (CASEVAC) and treatment that helicopters represented. A Royal Navy helicopter also carried out the first combat rescue when the pilot of a Hawker Sea Hawk fighter-bomber, Lt Stuart-Jervis, had to eject.

ABOVE: **A flight of six Bristol Sycamore helicopters transporting troops of 45 Commando Royal Marines during the assault on Port Said.**

International, mainly US, pressure forced a ceasefire, and the British and French forces ultimately withdrew. History judges the Anglo–French military action to have been pointless, but the campaign did showcase the remarkable versatility of the helicopter and its many military applications. It also persuaded British military chiefs that commando-carrying, dedicated amphibious warfare ships equipped only with helicopters were an essential component of the Fleet.

BELOW: **British helicopters were also used to transport Egyptian wounded to hospital ships for treatment. A Westland Whirlwind prepares to depart Port Said.**

LEFT: **Not all helicopter development in Europe was successful. This strap-on helicopter was designed by Frenchman, George Sablie.** ABOVE: **The Aerotécnica AC-12, designed by Jean Cantinieau in France, had a distinctive spine above the cockpit, which carried the engine ahead of the rotor assembly. The type was first flown in 1954, and 12 machines were delivered to the Spanish Air Force as the EC-XZ-2.**

European helicopter development

Helicopters are often thought of as an US invention, but although many European companies began by licence-building US machines, technical vision and innovation was also very evident in Europe.

British helicopter development made little significant progress until 1944 when the Bristol Aeroplane Company established a Helicopter Division, later to become Bristol Helicopters. The company's first helicopter was the Sycamore, which was in some ways more technically advanced than the early Sikorsky machines, and won export orders. The Bristol team then went on to develop the twin-engine tandem rotor Bristol Type 173 that became the Belvedere. Meanwhile, Westland was manufacturing a modified Sikorsky S-51 under licence as the Dragonfly for the British military and export customers. In 1960, Bristol Helicopters was absorbed by Westland, which was already established as the main British helicopter manufacturer. Westland went on to manufacture other Sikorsky types through to the Sea King.

Westand soon began to co-operate with French aerospace companies, and this led to the hugely successful Gazelle and later the Lynx. Italian company Agusta and Westland joined together to develop the design that ultimately produced the versatile and successful Merlin. Today the companies have merged as AgustaWestland.

Two French companies, Sud-Est and Sud-Ouest, developed a range of exciting helicopter projects in the late 1940s and early 1950s. Sud-Ouest built the tipjet Djinn, while Sud-Est developed the record-breaking Alouette series of helicopters. Sud-Ouest and Sud-Est were later merged into Sud-Aviation, which itself became part of Aérospatiale. It was this team that developed the Puma that is still in service today.

In 1992, Aérospatiale joined with the German helicopter manufacturer Messerchmitt-Bölkow-Blohm (MBB) to create Eurocopter Holdings. Both companies were already participating in the four-nation NH Industries NH90 military helicopter project, in partnership with Agusta and Fokker.

ABOVE: **The French-built Sud-Ouest SO-1120 Ariel III. Note the single fin and rudder tail assembly. Additional directional control was provided by turbine exhaust gases blown over the fin.**

ABOVE: **The Agusta A.101 was a large transport helicopter built only as a prototype and was first flown on October 19, 1964. Flight testing and development continued until 1971 when the project was abandoned.**

LEFT: **The Nord 500 was a single-seat tilting-duct research aircraft. Two were built in France, the second flying in 1968. Two five-blade ducted propellers and the wing on which they were mounted could be tilted horizontally.**

The post-war German helicopter industry had unlikely origins. In 1956, a civil engineering company moved into aviation, and formed Bölkow-Entwicklungen KG. In 1957, the company decided to investigate the possibility of designing its own helicopter trainer to capitalize on the then greatly expanding market. The P102 prototype was built, and was then produced as the Bo102. It was constructed as a training airframe that was never intended to fly, and it sold well. Having built one flying example, the team then developed more marketable helicopters, including the ultimately very successful Bo105 that featured an innovative rigid rotor system. After a succession of mergers, including one with the famed Messerschmitt company, the operation was finally known as Messerschmitt-Bölkow-Blohm (MBB).

The Agusta company in Italy (now part of AgustaWestland) is the European helicopter company with the longest aviation lineage. It was founded in 1923 by Count Giovanni Agusta, who had flown his first aircraft in 1907 and built aircraft until World War II. Post-war, Agusta switched to motorcycle manufacture, and then from 1952 the company became involved with helicopters, building Bell, Sikorsky, Boeing and McDonnell Douglas products under licence. The company, however, had ambitions to design and build its own helicopters, and these have included the Agusta A109 and the A129 Mangusta anti-tank helicopter – the first attack helicopter to be designed and produced in Western Europe.

It was the end of the Cold War and the associated budgetary limitations and high development costs of new technology that encouraged many companies to amalgamate and produce specific aircraft, or to merge entirely. This is the only way that the European helicopter industry has a real chance of competing with US and Russian military helicopter manufacturers.

LEFT: **The Eurocopter AS 350 Ecureuil (Squirrel in RAF and AAC use) and AS 355 Ecureuil 2 were designed in the early 1970s as a replacement for the Aérospatiale Alouette II (lark).**

ABOVE: **The Bristol Type 175 Belvedere was operated by No.66 Squadron in Malaya from June 1962, providing heavy-lift support for British forces on ground operations.**

The Malayan Emergency

During World War II, Malayan communist anti-Japanese guerrillas had been supported by the British, and after the end of the war these same guerrillas attempted to seize power in Malaya. Britain's response was to declare a State of Emergency in June 1948, and rush reinforcements to the country. At any given time, some 3,000 guerrillas were fighting against a combined force of around 350,000 troops, police and militia. The emergency continued for 12 years, and since much of the fighting took place in deep jungle, progress was slow. The introduction of helicopters made a huge difference to the British military effort. Entering service as a CASEVAC asset, the helicopters were soon in use transporting troops quickly to launch surprise attacks. This new type of air mobility enabled the British military to fight a different kind of war.

Malaya was one of the first operational deployments of a Royal Air Force helicopter unit – the Far East Air Force (FEAF) Casualty Evacuation Flight (CEF). This was formed in response to a request made to the Chiefs of Staff in March 1949 for the provision of helicopters to evacuate any casualties resulting from a planned increase in operations against terrorists in the more remote jungle areas. As a result, the CEF was formed in April 1950, and equipped with three (only three were ever built) Westland Dragonfly Mk2 helicopters based at Changi, Singapore, from May 1950. Their first CASEVAC operation was to recover a British soldier who had been shot during an ambush and flown from a water-logged airstrip at Segamat on June 14. The first CASEVAC operation from a jungle clearing followed just five days later.

The CEF was re-equipped with the more capable Dragonfly HC.4 which, like the Mk2, was a dedicated CASEVAC version built for the RAF and fitted with all-metal rotor blades.

LEFT: **Westland Whirlwind HAR.4s of No.155 Squadron near Kuala Lumpur airfield. During the Malayan emergency, the squadron was primarily deployed on CASEVAC duties.**

ABOVE: **Men of 22 Special Air Service Regiment (SAS) guiding a Bristol Sycamore HR.14 of No.194 Squadron to land in a jungle clearing at Ula Langat, near Kuala Lumpur, November 1957.** RIGHT: **Troops of the Special Air Service deploying from a Whirlwind HAS Mk22 of the Royal Navy.**

The helicopters, of course, were also used for purely military missions, including reconnaissance, the tactical movement of British troops, and delivery of military supplies and equipment.

The success of the CEF led to it being elevated to full squadron status as No.194 in February 1953. The Dragonfly helicopters were in constant demand, and from early June 1953 were supplemented by aircraft from No.848 Naval Air Squadron (NAS) equipped with early Sikorsky-built S-55 Whirlwind. No.194 Squadron operated the Dragonfly Mk4 until October 1954, when it began to receive the Bristol Sycamore Mk14, and finally phased out the last Dragonfly in July 1956. In April 1957, it was reported that these machines had flown 10,000 hours, one aircraft alone having flown for 1,000 hours.

In October 1954, helicopter operations were further strengthened by the introduction of the Westland Whirlwind Mk4 to equip No.155 Squadron formed at Kuala Lumpur airfield, which had now become the main helicopter operations base. By the end of 1956, No.155 and No.194 were equipped with 17 Whirlwind and 14 Sycamore helicopters, but a reduced threat and continuing problems with the Whirlwind led to the only helicopters operating in Malaya being the Sycamores of No.194. In February and April 1959, two fatal accidents involving Sycamore helicopters led to the remainder of the force in Malaya being grounded.

On June 3, 1959, No.194 and No.155 Squadrons were disbanded, and No.110 Squadron re-formed at Butterworth with the remaining Whirlwinds from No.155. On July 31, 1960, after the squadron was re-equipped with the Bristol Sycamore Mk14, the Malayan Emergency was declared to be at an end. The celebratory flypast over Kuala Lumpur included three Sycamore Mk14s.

The British helicopter force in Malaya pioneered most of the short-range military helicopter transport roles, and from June 1950 to July 1960 these RAF and RN helicopters carried out 4,759 evacuations, moved 127,425 troops, 17,865 passengers and 1,226,034kg/2,699,327lb of cargo. These figures are all the more remarkable when one remembers that the first-generation helicopters were only capable of transporting five soldiers and a maximum of 363kg/800lb at a time.

ABOVE: **A Westland Whirlwind being recovered from a swamp after a forced landing in a jungle area near Kuala Lumpur.**

The Vietnam War

While the Korean War (1950–53) was to prove the military value of the helicopter, Vietnam was the first war that could not have been waged without the type. From the first US helicopter combat missions captured by television news crews to the chaos of the final US withdrawal from Saigon rooftops in 1975, Vietnam was the first true helicopter war.

US forces faced an unusually complex challenge in South-east Asia. While fighting to support the Republic of Viet Nam in a battle against a communist-backed insurgency, US forces had to establish a very modern fighting force in an underdeveloped environment with challenging terrain while facing Viet Cong (VC) guerrilla tactics and conducting conventional operations against well-trained regular units of the Viet Nam People's Army (North Vietnamese Army). Helicopters were vital to US operations, and thus began the development of the US air mobility force.

Among the first types operated by the US Army in South Vietnam were Bell HU-1A Iroquois "Hueys" that arrived in April 1962, before the US was officially involved in the war. These supported the Army of the Republic of South Viet Nam (ARVN), but were flown by US crews. In October that year, the first armed Hueys, equipped with 2.75in rockets and 0.30in machine-guns, began operations as gunships for escort missions in support of the transport helicopters.

Piasecki H-21 Shawnee helicopters were used to ferry ARVN troops into battle against VC forces in 1962. When troop-carrying helicopters first appeared in action,

ABOVE: **The Bell UH-1A Iroquois Huey is an iconic symbol of the Vietnam War due to the presence of the type in great numbers.**

BELOW: **A US Marine Corps Sikorsky HUS-1 flying over South Vietnam. The type was extremely vulnerable to ground fire.**

LEFT: **A Bell UH-1A Iroquois fitted with chemical spraying equipment. Operations included the eradication of mosquitoes, destroying Viet Cong farm crops, and tree defoliation.**

the VC troops fled, but at the Battle of Ap Bac in 1962 they stood firm and shot down five US helicopters, while damaging many others. Later Communist combat training manuals were to show the best techniques for shooting down helicopters by using small arms as a form of improvised anti-aircraft fire, shooting ahead of the target to allow for forward speed to increase the chance of hitting the aircraft. The manuals also detailed the most vulnerable parts of the machine (tail rotor, rotor head and engine), where the most devastating mechanical damage could be inflicted.

The Bell UH-1 Iroquois was the workhorse of the war, and according to the Vietnam Helicopter Pilots Association (VHPA), those in US Army service flew 7,531,955 hours between October 1966 and 1975. Operations with the Bell AH-1G Cobra accumulated 1,110,716 flight hours in the war. As a result, some sources, including the VHPA, assert that these two types have more combat flight time than any other aircraft in the history of warfare, if one includes exposure to actual hostile fire. For example, Allied bombers in World War II were certainly flown on long missions, but were not exposed to enemy fire for the entire time. In Vietnam, helicopters were rarely flown above 458m/1,500ft, and thus were always at risk from enemy ground fire.

LEFT: **South Vietnamese paratroopers running to board two Piasecki CH-21 Shawnee helicopters for a mission in the Mekong Delta. A large number of the type were shot down by enemy ground fire.**

ABOVE: **The HH-3E Jolly Green Giant was used for combat rescue, and featured protective armour, self-sealing fuel tanks, a retractable inflight refuelling probe and a high-speed hoist.**

The Bell UH-1 was upgraded and improved, based on combat experience gained, and was instrumental in enabling the US military to develop airmobile warfare tactics. A typical air assault mission would see UH-1 helicopters dropping infantry deep inside enemy territory while gunships would escort an attack force of up to 100 transport helicopters. In a matter of minutes, battalions of US troops could be inserted behind enemy lines, ready for combat. The troops were deployed without having to fight for positions, or being too dispersed along a wide battlefront, and could be withdrawn as required. This meant that ground would not be taken, but it allowed strategists to deploy military assets over large areas and engage the enemy at designated places.

The statistics detailing the use of helicopters in the war are staggering. A total of 2,197 US helicopter pilots and 2,717 crew members were killed from all services including those from Air America, an "airline" operated by the Central Intelligence Agency (CIA). Some 500,000 MEDEVAC missions were flown and over 900,000 casualties were airlifted. As a result of these operations, the average time lapse between battlefront to hospitalization was less than one hour, resulting in significantly improved survival rates for casualties.

Militarily, the helicopter allowed US forces unprecedented mobility. Without the aircraft it would have required three times as many troops to secure the 1,287km/800-mile border with Cambodia and Laos. About 12,000 helicopters were deployed to Vietnam with a peak strength of some 4,000 in 1970, of which around 2,600 were Bell UH-1s. A total of 5,086 helicopters were lost in service over Vietnam, including over 1,000 in 1968.

Mention must also be made of Royal Australian Air Force (RAAF) and Royal Australian Navy (RAN) helicopter operations in South Vietnam. Although small in scale

RIGHT: **Sikorsky CH-53 from HMH (Marine Heavy Helicopter Squadron) 462 on a mountainside base in Vietnam in 1968. At this stage in the war, the USMC squadron was tasked with the tactical retrieval of downed aircraft and movement of heavy equipment.**

ABOVE: **Helicopters enabled US forces to operate from a variety of locations, including smaller warships.** RIGHT: **Inflight refuelling from aircraft such the Lockheed C-130P Hercules enabled Combat Search and Rescue (CSAR) helicopters to range far and wide over South-east Asia.**

compared to US operations, the overall contribution of these two services in the conflict is often and quite wrongly overlooked by historians outside Australia.

The RAN formed the Royal Australian Navy Helicopter Flight Vietnam (RANHFV), and the first element arrived in October 1967. Trained for anti-submarine warfare, the crews had to learn how to drop and recover troops in high-risk combat areas. The RANHFV was integrated with the US 135th Assault Helicopter Company tasked with the tactical movement of combat troops, supplies and equipment. During a one-year operational tour, RAN crews commonly logged a combined total of between 9,000 and 12,000 flying hours. The RANHFV ceased offensive operations on June 8, 1971, and by the time the unit left the country, hundreds of offensive operations had been flown. A total five personnel were killed and 22 wounded in action from a complement of 200.

An element of No.9 Squadron RAAF had arrived in Vietnam in June 1966, complementing to the RAAF's fixed-wing commitment of military transports and, later, bombers. Equipped with UH-1Bs, the RAAF helicopters transported infantry, equipment and supplies, and dropped propaganda leaflets over enemy territory. The aircraft were used to spray their base to eradicate mosquitoes but, more offensively, to spray chemicals on Viet Cong farmland to interrupt the enemy food supply. In 1967, after re-equipping with the more capable UH-1H, No.9 worked more closely with its RAN counterparts, transporting troops to and from patrols, and undertaking battlefield casualty evacuations. By the time No.9 finally left Vietnam in December 1971, six personnel had been killed in action.

LEFT: **The Sikorsky HUS-1 was widely used in Vietnam by the USMC, and although the US Army operated the CH-34 Choctaw, the type was not used in the war.**

LEFT: **A US Navy Sikorsky HO3S-1 of Helicopter Utility Squadron HU-1 taking off from the battleship USS *New Jersey* (BB-62) off Korea on April 14, 1953.**

Helicopters at sea

As the helicopter was developed for military use, its value as a naval, particularly anti-submarine weapons platform, was exploited specifically. Planners could see the value of an aircraft that could take off vertically from a relatively small space while carrying troops or weapons. Initially, only aircraft carriers could cope with the large weapons and equipment-laden helicopters such as the Sikorsky S-58 but, after trial operations, it was confirmed that smaller warships – mainly frigates tasked with defending other ships, if modified or ideally designed for purpose from the outset – could operate these vital aircraft effectively. By operating an ASW helicopter, these ships could extend a navy's anti-submarine capability far beyond the fleet the frigates were there to protect from enemy attack.

During the Cold War, Soviet submarines were considered to be a major threat as they had the ability to attack NATO fighting ships, just as the German Navy U-boats had for a time wreaked havoc with Allied shipping in the Atlantic during World War II. As missile technology improved and submarines were developed to fire nuclear missiles at the West while submerged, it became even more important to be able to detect and destroy enemy submarines in war.

The Royal Navy has always been at the forefront of helicopter-borne anti-submarine capability, and the RN's Type 81 Tribal class of frigate commissioned between 1961 and 1967 were the first designed from the outset to operate a helicopter. On the Tribal class, this would be a single Westland

RIGHT: **A Kamov Ka-27PS (NATO identifier Helix) from the Ukranian Navy landing on the flight deck of USS *Taylor* (FFG-50) in the Black Sea during Exercise Sea Breeze, 2010.**

LEFT: **An AgustaWestland EH-101 Merlin on the deck of HMS *Iron Duke* (F234).**
ABOVE: **A South African-built Atlas Oryx M2 landing on the deck of USS *Swift* (HSV-2).**

Wasp equipped to carry two homing torpedoes, but could also be armed with SS.11 wire-guided missiles to attack small surface vessels.

The Type 81-class ships were complemented by the Leander-class frigates, the last of 26 of which entered service in 1971. These ships had a helicopter hangar for the anti-submarine warfare Westland Wasp.

From the early 1960s, the RN also operated Type 12 frigates, the Whitby class, designed as an ocean-going convoy escort. The Type 12s were used as fast fleet anti-submarine warfare escorts and were also equipped with a Westland Wasp.

On April 25, 1982, the Argentine submarine *Santa Fe* was spotted and attacked first by a Royal Navy Westland Wessex from the destroyer HMS *Antrim* (D18). HMS *Plymouth* (F126), a Type 12, then launched its Wasp HAS Mk1 while HMS *Brilliant* (F92), which carried two Lynx helicopters, launched one of its helicopters to join the attack. Two other Wasps from the RN Antarctic ice patrol ship HMS *Endurance* (A171) also took part in the attack. Anti-shipping missiles were fired at the submarine from the helicopters, causing damage that forced the vessel to be withdrawn. Two interesting historical facts to note are that the Wasp had already been retired from RN service when it was brought back into service for the Falklands War. Also, when the RN sailed south to retake the Falklands, it was carrying nuclear depth charges that would have been helicopter-dropped if required.

Today the Royal Navy operates Type 23 frigates, first conceived in the late 1970s to specifically counter Soviet nuclear submarines operating in the North Atlantic. First commissioned in 1989 in the anti-submarine role, these ships have become multi-purpose fighting ships reflecting both the demands of the world today and ever-shrinking defence budgets. One of these ships is the HMS *Iron Duke* (F234) which operates Lynx Mk3 or Mk8 helicopters in the anti-submarine role, although these, like the ship, can be used for other missions. Other frigates in the class operate the Merlin helicopter in place of Lynx. The RN believes that this weapons system has re-affirmed its reputation as a leader in anti-submarine warfare.

LEFT: **Civil-registered Super Puma helicopters operate from Military Sealift Command (MSC) transport stores ships, and are used for Vertical Replenishment (VertRep) of other US Navy vessels.**

The helicopter cockpit

The helicopter is a the world's most versatile flying machine, but very complex and difficult to fly – a pilot must be able to coordinate in three dimensions and must constantly use both hands and feet simultaneously to keep the aircraft under control.

The cockpit of a helicopter is at first glance similar to that of a fixed-wing aircraft, but closer inspection reveals significant differences. Helicopters do not have the same control wheels or columns, and usually no throttle controls. The other significant point to note is that in most helicopters the pilot-in-command flies the machine from the right-hand seat, whereas on the flight deck of a fixed-wing aircraft the captain flies the aircraft from the left-hand seat.

In the cockpit, the cyclic stick (control column) is operated to tilt the main rotor to raise or lower the nose and to make the aircraft bank to the left or right. The collective pitch control is positioned to the left of the pilot's seat, and is operated to change the pitch angle of the main rotor blades to decrease or increase lift for take-off, landing and flight altitude.

A helicopter does have a conventional rudder for directional control, but is fitted with an anti-torque tail rotor. The pedals operated by the pilot's feet are not unlike those in the cockpit of a fixed-wing aircraft, and are positioned in the same place. The pedals control the pitch of the tail rotor blades to provide

TOP: **The AW101 Merlin in service is equipped with high-definition colour display screens.** ABOVE: **The much simpler all-analogue dial instrument panel of the Sikorsky S-51.**

directional control while the machine hovers and is in forward flight. Helicopter and fixed-wing aircraft pilots have to receive the same technical information, such as airspeed, attitude (in relation to the horizon), altitude, compass heading and rate of climb and descent, to fly safely and efficiently. Engine performance, including fuel and oil consumption, is vital to the operation of both types of aircraft. The helicopter is also equipped with a torque indicator to show the amount of engine power being applied to the main rotor blades.

Analogue (dial-type) instruments were a standard fitment in all cockpits for many years, but many of the types now in service are fitted with all-glass cockpits. All-important flight data, navigation and weapons information is displayed on small, flat Liquid Crystal Display or Video Display Unit (LCD or VDU) screens. Most military helicopters have basic analogue flight instruments as a stand-by in the event of an electronics systems failure. Other equipment in the cockpit of modern military helicopter is fitted to provide communications and navigation information, with specialized warning equipment for defence and targeting for offensive weaponry.

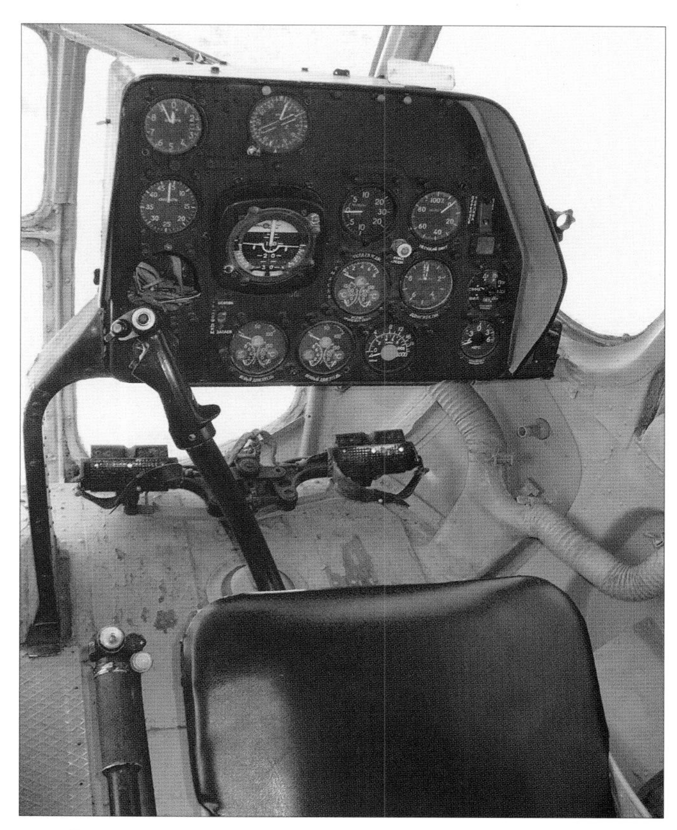

ABOVE: **The very basic instrument panel of the Mil Mi-8. Note the collective, bottom left, the cyclic in front of the seat, and the foot pedal control bar.**

BELOW: **The AgustaWestland AW139 has a Honeywell Primus Epic modular/integrated glass cockpit, offered to customers in four different mission-optimized configurations. Maintenance personnel may use the cockpit displays or a laptop computer to perform aircraft rigging, sensor calibration and avionics systems diagnostics.**

LEFT: **The pilot of an AgustaWestland Merlin in Royal Navy service carrying out his pre-flight checks. The complex rotor systems fitted on modern military helicopters can survive much greater damage than those of earlier machines.**

Rotor systems

Autogyro pioneer Juan de la Cierva investigated, developed and tested many of the fundamentals of the rotor (the system of rotating aerofoils) that lifts and keeps a helicopter off the ground. A helicopter rotor is powered by the engine, through the transmission (the mechanism transmitting mechanical energy) to the rotating mast (a metal shaft) extending up from and driven by the transmission. At the top of this mast is the rotor hub, the attachment point for the individual rotor blades.

The pitch of main rotor blades can be varied cyclically throughout its rotation to control the direction of rotor thrust and select the part of the rotor disc where the most thrust will be developed – for example front, side and back. The collective pitch varies the magnitude of rotor thrust, increasing or decreasing thrust over the entire rotor disc at the same time. The differences in blade pitch are controlled by tilting and/or raising or lowering what is known as the "swash plate" with the flight controls.

The swash plate generally consists of two closely spaced concentric plates separated by bearings, the top one of which rotates with the mast, while the other does not. The rotating plate is also connected to individual rotor blades, and the non-rotating plate is connected to links which are moved by the pilot – the collective and cyclic controls. The swash plate can move vertically and tilt. Through shifting and tilting, the non-rotating plate controls the rotating plate, which in turn controls the pitch of each individual rotor blade.

BELOW LEFT: **The Cierva W.11 Air Horse was the largest helicopter in the world when it first flew in 1948. The three rotors mounted on outriggers were driven by a single engine mounted inside the fuselage.**
BELOW: **The Yakovlev Yak-24 had a tandem rotor layout, not typical for Soviet helicopters. The engines were linked so that each could drive one or both rotors, but this caused severe vibration in the airframe.**

LEFT: **The rotor head of a Sikorsky S-92. This helicopter features an active vibration control system that provides comfortable flight and acoustic levels. This system also extends airframe life by reducing fatigue loads on the airframe.**

Most helicopters maintain a constant main rotor speed, so only the rotor blade angle of attack is used to adjust thrust. The mighty Mil Mi-6, for example, has a main rotor turning at a very slow 120rpm during flight.

Helicopter main rotor systems are classified depending on how the main rotor blades are attached and move relative to the rotor hub. Essentially there are three basic types: fully articulated, rigid and semi-rigid.

Cierva developed the fully articulated rotor for systems with three or more blades. Each rotor blade is attached to the rotor hub through a series of hinges which allow each blade to move independently. The horizontal hinge, called the flapping hinge, enables the blade to move up and down, and this movement (flapping) is designed to compensate for dissimilar lift and ensures a steady, level take-off. The flapping hinge can be at varying distances from the hub, and there may be more than one flapping hinge. A vertical hinge, called the lead-lag or drag hinge, allows the rotor blade to move back and forth. The purpose of the drag hinge and dampers is to compensate for acceleration and deceleration and, by changing the pitch angle of the blades, the thrust and direction of the main rotor disc can be controlled.

A rigid rotor is usually a hingeless rotor system where blades are flexibly attached to the rotor hub. Rigid rotors are either of the feathering or flapping type. In a flapping rigid-rotor system, each blade flaps, drags and feathers (depending on the design) about flexible sections of the root. The flapping rigid-rotor system is mechanically simpler than the fully articulated rotor system.

LEFT: **The Kaman HH-43 Huskie had an unusual intermeshing contra-rotating twin-rotor system.** ABOVE: **The Russian-built Kamov Ka-25 has folding three-blade coaxial rotors.**

ABOVE: **The McDonnell Douglas MD 520N is fitted with the NOTAR anti-torque system, which increases operational safety and reduces external noise.**

Loads from flapping and lead/lag forces are accommodated by bending rather than through hinges. By flexing, the blades compensate for the forces that previously required rugged hinges. The result is a rotor system that has less of a delay in control response because the rotor has much less oscillation.

The semi-rigid rotor is also known as a teetering or seesaw rotor. This system is always composed of two blades (such as on a Bell UH-1) which meet just under a common flapping, or teetering, hinge at the rotor shaft. This allows the blades to flap together in opposite motions, like a seesaw. The semi-rigid description comes from the fact that it does not have a lead-lag hinge. The rotor system is rigid in-plane, because the blades are not free to lead and lag, but they are not rigid in the flapping plane (through the use of a teetering hinge). The rotor is therefore not rigid, but neither is it fully articulated, so it is termed "semi-rigid".

Most helicopters have a single main rotor, but a separate rotor is also required, rotating in the vertical plane to overcome torque, the effect that causes the body of the helicopter which is attached to the engine to always turn in the opposite direction of the rotor. To eliminate this effect, some type of anti-torque control must be designed into the helicopter to allow the aircraft to maintain a heading and to provide yaw control. The three most common types of anti-torque controls are the tail rotor, ducted fan and NOTAR (no tail rotor).

The tail rotor is a small propeller mounted to rotate at or near the vertical at the end of the tail boom of a typical single-rotor helicopter. The tail rotor's position and distance from the aircraft's centre of gravity allows it to develop thrust in the opposite direction from the rotation of the main rotor, and thereby balances the torque effect. The tail rotor is far less complex than the main rotor, as it requires only collective changes in pitch to vary thrust. It is controlled by the pilot using foot pedals, which also provide directional control by

LEFT: **The semi-rigid rotor system is used on helicopters such as the Bell UH-1 Iroquois and Bell AH-1 Cobra.**

allowing the pilot to rotate the entire helicopter around its vertical axis and, in doing so, change the direction in which the machine is pointed.

Originally conceived by Sud Aviation in France, a ducted fan can be used in place of a tail rotor and built into the aircraft tail with the fan housing being part of the aircraft, as in the case of the Gazelle. A ducted fan can typically have eight to 18 blades, compared to the two to four of a standard type of tail rotor, and can have irregular angular spacing, allowing it to run at lower noise levels. Although heavier and more expensive to build, the ducted fan is less prone to foreign object damage or tail strike as the aircraft is taking off or landing, and is less likely to cause injury on the ground. Fenestron and Fantail are trademarked names for individual manufacturers' ducted fan designs.

The third anti-torque option is NOTAR, an acronym trademarked by Hughes Helicopters for NO TAil Rotor. The use of directed air to provide anti-torque control was tested as early as 1945 on the British-built Cierva W.9, but Hughes (later McDonnell Douglas) engineers revisited the concept in the mid-1970s. Although the principle took time to refine, it is quite simple, and provides directional control in the same way a wing develops lift. A variable pitch fan driven by the main rotor transmission forces low-pressure air through two slots on the right-hand side of the tail boom. This causes the downwash from the helicopter's main rotor to hug the tail boom, producing lift, and directional control which is supplemented by a direct jet thruster and vertical stabilizers. Three McDonnell Douglas production helicopters currently use the NOTAR system: the MD 520N (a NOTAR variant of the Hughes/MD500 series), the MD 600N (a larger version of the MD 520N), and the MD Explorer.

ABOVE: **The conventional tail rotor on an Aérospatiale Puma generates thrust to counteract the main rotor torque and directional control.**

LEFT: **Fenestron is the trademarked name of the integrated ducted fan assembly fitted to the Eurocopter EC120B. Comparing the two systems, the enclosed fan is much safer for ground operation, and is also quieter.**

LEFT: **Vietnam, 1966. Ease of engine maintenance is a key design consideration. The Sikorsky S-58 had unique clamshell covers that gave easy access to the engine. Note the engine is mounted at an angle.**

Helicopter engines

A helicopter's size, performance and its applications are all determined by the number, size and type of engine used. Without the right powerplant a helicopter is simply a large ornament, and the correct engine is as vital as the rotor that keeps the aircraft in the air. Incredibly powerful steam engines were available in the 19th century, but the power from even, by necessity, the smallest, lightest examples was not sufficient to allow for manned helicopter flight.

It should be remembered that although Forlanini was astounding people with his contra-rotating steam-powered model in 1877, it was just that – an experimental model and a long way from carrying a man off the ground. Just as it was with fixed-wing aircraft, it was the arrival of the internal combustion engine in the late 19th century that started to turn the dream of manned helicopter flight into reality. Increasingly powerful, more compact engines began to be developed which, it was hoped, could produce enough power to lift their own weight and a helicopter and pilot off the ground. Early helicopter designs used custom-built engines or rotary engines that had been designed for fixed-wing aircraft, but more powerful motor car and radial engines soon replaced these in the plans of aspiring helicopter designers. Even so, what held back helicopter development more than anything else was the overall lack of power required to overcome the machine's weight in vertical flight.

It was when the compact, horizontally opposed flat engine was developed that helicopter designers found a lightweight powerplant that could be readily adapted to small experimental helicopters. As soon as the helicopter design was perfected, then the designers, on a quest for bigger and more capable helicopters, returned to the trusted radial engine. The Wright Cyclone R-1820 was a development of an engine that first ran in 1925. The engine was used to power wartime fixed-wing

LEFT: **One of two General Electric T64-GE-7 turboshaft engines on a Sikorsky CH-53G Super Stallion.**

LEFT: **The Aérospatiale Alouette II (lark) was the first production helicopter powered by a turboshaft engine instead of a piston engine, which was considerably heavier.**

aircraft, including the Boeing B-17 Flying Fortress bombers and even tanks such as the famous Sherman. A Wright Cyclone R-1820 mounted diagonally below the cockpit also powered the early versions of the Sikorsky S-58/H-34 Choctaw.

In the UK, the Alvis Leonides, first run in 1936, was chosen to power the Westland Dragonfly by direct drive immediately beneath the rotors. Later versions of the Westland Whirlwind were also powered by inclined-drive versions of the same engine. These helicopters were later re-engined with the lighter and more powerful Bristol Siddeley Gnome turboshaft. The arrival of the jet engine immediately opened a wealth of possibilities for designers, once it was modified and developed into a turboshaft which optimized shaft power rather than jet thrust. Crucially, the turboshaft engine was able to be scaled to the size of the helicopter being designed, so that virtually all but the lightest of helicopters are powered by turbine engines today.

In December 1951, the Kaman K-225, predecessor to the HH-23 Huskie, had been the first-ever turboshaft-powered helicopter to fly. Kaman replaced the piston engine with a Boeing 50-2 turboshaft to demonstrate the potential of jet-powered helicopters to the US Navy.

German wartime jet development expertise was harnessed by the Lycoming company who, in 1951, appointed Anselm Franz. He had led the design of the Junkers Jumo 004, the world's first turbojet engine, to enter production, which then went on to power the Messerschmitt Me 262 jet fighter and Arado Ar 234 bombers in combat. Franz was taken on to develop the T35 turboshaft for helicopters. Around 20,000 of these engines were made in turboshaft and turboprop versions. The turboshaft was used to power the Bell AH-1 Cobra, UH-1 Iroquois and Kaman H-43B Huskie.

French manufacturers explored tipjet propulsion – the 1949 SNCASO 1100 Ariel I used a piston engine to drive a compressor that pumped air under pressure to tip burners to spin the rotors. It was the Sud-Ouest Djinn (genie), with simpler cold tip jets that expelled compressed air without a combustion chamber, that proved to be a success. The French Alouette II (lark) fitted with a Turboméca engine became the first production helicopter to be turboshaft-powered.

The jet engine made helicopters fly better, faster and for longer, and removed the need to carry dangerous, low-flashpoint, gasoline-based fuels. Fewer moving parts meant engine failure was a rare occurrence compared to piston-powered machines. A turboshaft-powered helicopter can usually be ready to lift off within just a minute of engine start.

ABOVE: **The Allison/Rolls-Royce Model 250 turboshaft has been used to power helicopters, including the Bo105 and Bell OH-58 Kiowa.**

ABOVE: **The Lycoming T55 turboshaft engine was first developed over 50 years ago, and was used on both helicopters and as a turboprop on fixed-wing aircraft.**

Boeing CH-47 Chinook

In the late 1950s, the Vertol Aircraft Corporation began development of the Model 107 (V-107) as a replacement for Piasecki H-21 Shawnee and Sikorsky H-34 Choctaw piston-engine helicopters. The US Army awarded a contract to Vertol in 1958 to produce the YHC-1A, but decided that the machine did not have an adequate performance for heavy-lift operations. Boeing acquired Vertol in 1960 and produced the larger Model 114 as the YHC-1B. On entering service the type was designated CH-47 and named Chinook. The US Army deployed the type to South Vietnam in February 1966, and it has remained in front-line service ever since.

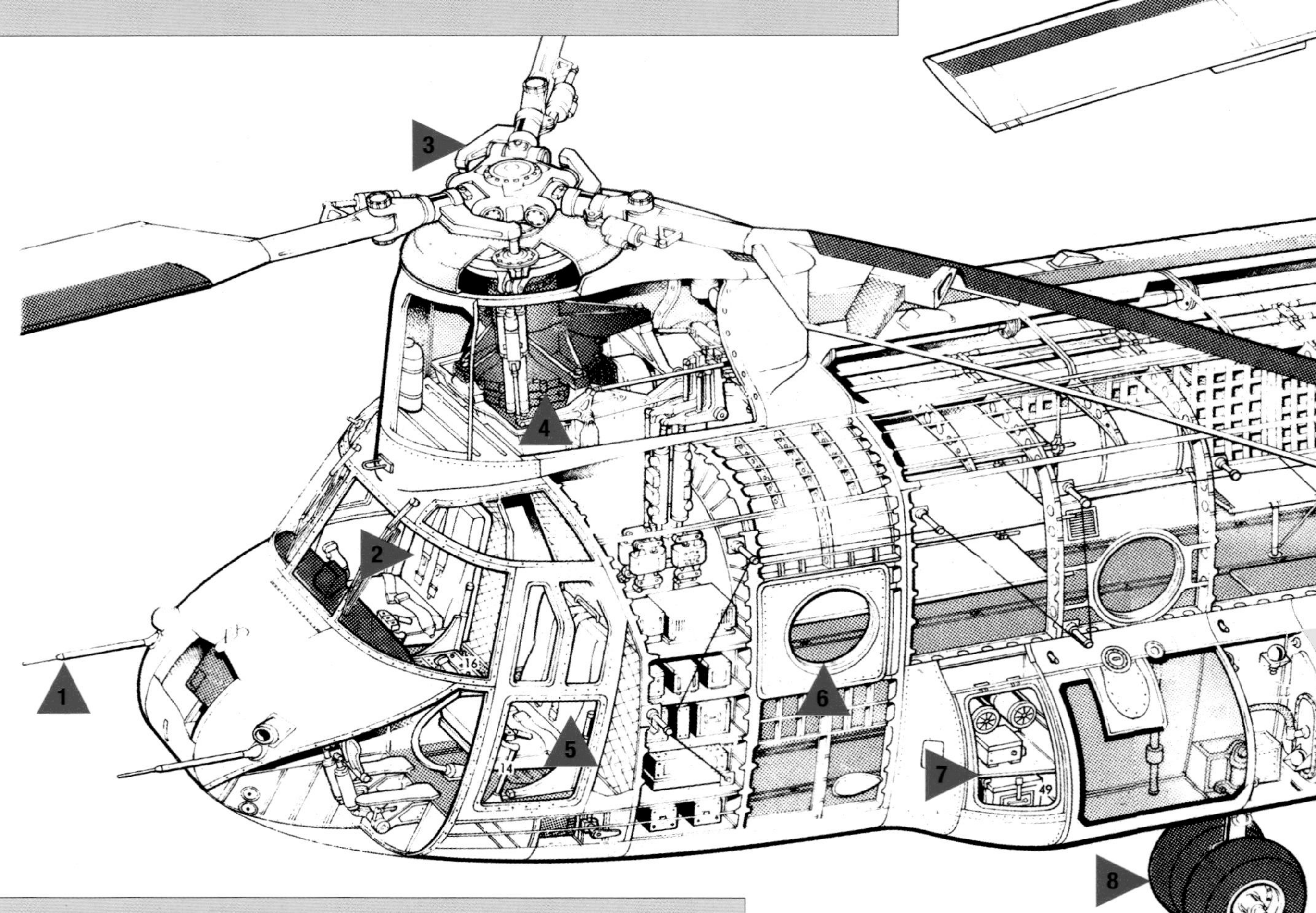

Key

1 Pitot head
2 Pilot's seat
3 Forward rotor hub
4 Forward transmission gearbox
5 Co-pilot/observer's seat
6 Emergency escape window
7 Electrical equipment bay
8 Main landing wheels
9 Main fuel tank
10 Rear landing wheel
11 Rear door
12 Vehicle loading ramps
13 Solar T-62T-28 auxiliary power unit
14 Rear rotor hub
15 Rear rotor driveshaft
16 Avco Lycoming T-55-L-712 turboshaft engine
17 Transmission combining gearbox
18 Main transmission shaft
19 Engine intake dust screen
20 Rotor blade

ABOVE: **When this large and powerful twin-engine, tandem-rotor medium-lift transport was first flown, it proved to be faster than a number of contemporary smaller utility, and even attack, helicopters. Around 1,200 have been built, and new examples are still being produced.**

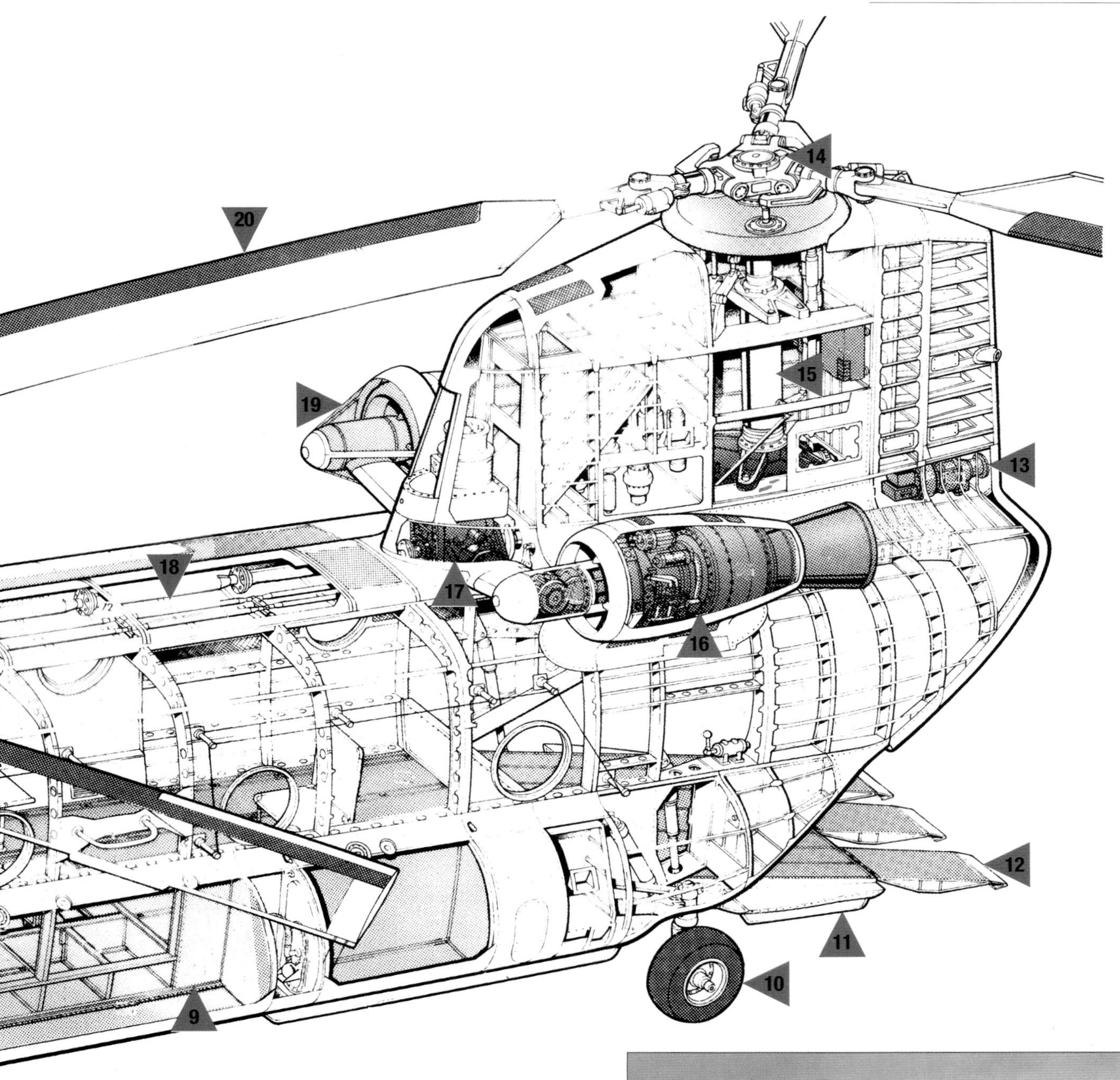

ABOVE: **The Chinook's powerful twin engines are mounted on the sides of the rear rotor mast, and drive both main rotors.**

ABOVE: **The CH-47 has proved itself to be a medium-lift workhorse in all environments, and will remain in service for many years to come.**

LEFT: **An image captured from a night vision device shows Sikorsky MH-53 Pave Low helicopters from the 20th Expeditionary Special Operations Squadron in Iraq departing on the last-ever combat mission before the type retired from service.**

Night Vision Technology (NVT)

One of the technical innovations that gives many military helicopters a vital edge over opponents and allows surprise night-time attacks while operating unmolested under cover of darkness is Night Vision Technology (NVT). During Operation Desert Storm (the first Gulf War), night vision-equipped attack helicopters enabled Coalition forces to launch repeated and undetected stand-off (from a distance) attacks on Iraqi tanks and military installations to reduce the threat to ground troops involved in the ground war.

Night vision equipment can enhance any available light by a factor of 10,000, essentially turning night into day even on a cloudy moonless night. This is especially useful when landing a helicopter at night in unfamiliar surroundings, where there might be obstructions such as trees or power lines. It also means that a pilot can land more confidently and in the minimum amount of time, which reduces exposure to enemy fire in the hover.

Night vision can work in two very different ways, depending on the technology applied – image enhancement or thermal imaging. Night vision imagery tends to be green, because the human eye can differentiate more shades of green than any other colour in the spectrum.

Image enhancement works by collecting the miniscule amounts of ambient light, including the lower portion of the infra-red light spectrum, that are present but are imperceptible to the human eye, and amplifying it to the point that aircrew can easily observe the image through an eyepiece or goggles attached to a helmet.

Thermal imaging operates by capturing the upper portion of the infra-red light spectrum emitted as heat by objects. Hotter objects, such as troops or recently run vehicles, emit more of this light than cooler objects such as buildings or trees.

In the AH-64 Apache, both the pilot and the gunner use night vision equipment for operations after dark. The night vision sensors work on the Forward Looking Infrared (FLIR) system, which detects the infra-red light released by heated objects. The pilot's night vision sensor is linked to a rotating turret on top of the helicopter's nose, while the gunner's night vision sensor is attached to a separate turret on the

LEFT: **Night vision equipment provides the user with a high-tech edge that could mean the difference between life or death in a combat environment.**

ABOVE: **Light-enhanced photography showing a US Air Force Sikorsky MH-53J Pave Low helicopter from the 6th Special Operations Wing during a mission over Afghanistan in support of Operation Enduring Freedom.** RIGHT: **Royal Marines of D Company, 40 Commando, are seen through Night Vision Goggles (NVG) as they depart HMS *Ark Royal* by Sea King HC4 helicopters, to take part in military operations on the Al Faw Peninsula in southern Iraq.**

underside of the nose. The lower turret also houses a standard video camera and telescopic sight, which the gunner uses during daylight operations.

A computer transmits the night vision or video picture to a small display unit built into each crew member's flight helmet. The display unit then projects the image on to a monocular (single) lens in front of the pilot's right eye. Infra-red sensors built into the cockpit track how the pilot positions his helmet and where his line of sight is, and then relays this information to the turret control system – a technique known as "slaving". Each pilot can aim the sensors on an AH-64 Apache by simply moving his head, and the system is designed to have a high rate of movement – 120 degrees per second – to accurately match head movement and ensure there is no time lag. The same technique is used for aiming the machine's devastating 30mm automatic cannon.

While first-generation NVT required the transition from looking through Night Vision Goggles (NVG) to looking down at the specially lit instrument panel with regular vision, the latest equipment offers a one-stop shop by superimposing all vital flight and instrument data in one view display on the flight helmet. This improves overall situational awareness and crew coordination, and speeds up reaction time for manoeuvring, targeting and for defensive actions, improving both crew and aircraft survivability.

LEFT: **A Sikorsky HH-60H Seahawk assigned to Helicopter Anti-Submarine Squadron 10 (HS-10), Expeditionary Sea Command Unit 1 (ESCU-1), preparing for a rescue mission to a stranded Taiwanese fishing vessel that has run aground on a reef near the Solomon Islands in the Pacific Ocean.**

LEFT: **The Boeing CH-47 Chinook twin-engine, tandem rotor helicopter has served with 17 nations since the 1970s. When the type first entered service, it was faster than many utility and attack helicopters.**

Military heavy-lift helicopters

Military transport helicopters are generally employed in preference to other forms of transport in situations where the cargo is required to be moved quickly, or because the destination is hard to access through other means, or the route over land or water to the destination presents threats to the cargo. Another reason is that the load to be carried might simply be an awkward shape or too heavy to transport easily by road or by fixed-wing aircraft. No other military asset can do what the helicopter, and especially the large helicopter, can do, as landing possibilities are almost limitless and in the situation where they cannot land, they can hover and set down underslung cargo or lower troops without even touching down.

When military forces have to be deployed quickly over any great distance to an area of operations, then large fixed-wing transport aircraft are used to get them to the area. Once there, it often falls to the large helicopter to move the troops and equipment to the fighting locations. Heavy-lift helicopters currently in service with military forces around the world include the Sikorsky CH-53 Sea Stallion and CH-53E Super Stallion, Boeing CH-47 Chinook, Mil Mi-26, and Aérospatiale Super Frélon. Capable of lifting up to 80 troops each and transporting small armoured vehicles (usually as slung loads but also internally in some cases), these helicopters have proved their worth in combat

RIGHT: **A Sikorsky CH-3C transporting a US Army Jeep-type vehicle mounted with a 106mm recoilless anti-tank rifle during Exercise Gold Fire in November, 1964.**

LEFT: **The Mil Mi-10 could normally carry 13,412kg/29,568lb internally, and up to 16,765Kg/36,960lb over a shorter distance. Alternatively, an underslung load of up to 8,941kg/19,712lb could be carried. The rear-facing gondola under the nose allowed a crewman to direct a lifting operation.**

numerous times. Mention should also be made of heavy-lift helicopters, including the Sikorsky CH-37 Mojave and the extraordinary Sikorsky CH-54 Tarhe, which can lift artillery pieces, trucks, medium tanks, other helicopters and boats, and be sent to recover downed aircraft.

The Boeing CH-47 Chinook has been operating around the world since it first flew in 1962. It has been regularly improved and has a cargo bay to accommodate loads well beyond the capacity of other helicopters, including heavy artillery weapons, light vehicles and even smaller helicopters. The heavy-lift use of one Royal Air Force Chinook helicopter illustrates how the type can help turn a battle. In 1982, during the Falklands War, a number of Special Air Service troops were dug in on a hill overlooking Port Stanley and were coming under artillery fire from Argentine forces. The only Chinook available at the time was tasked with delivering three 105mm howitzers to the troops at night. The three large guns were carried internally, and the ammunition was carried on pallets loaded in cargo nets slung under the helicopter.

The heavy-lift helicopter market has been dominated by companies in the USA and Russia by Mil, who specialized in developing large helicopters for the former Soviet Union. The Aérospatiale three-engine Super Frélon is, however, a notable exception, although it was developed with technical assistance from Sikorsky.

The closest that Britain came to developing a heavy-lift helicopter was the Westland Westminster, which was built and test-flown from 1958–60. The impressive machine could carry 6,359kg/14,000lb, and was described as the "largest twin-turbine mechanically driven single-rotor helicopter in the Western world". British helicopter industry mergers and government indifference led to the project being cancelled.

Mil has produced a number of large helicopters, but the largest was the Mil V-12, a twin-rotor helicopter that had a cabin more than 28m/92ft long that could carry a payload of 40,204kg/88,448lb. This machine was used to set many records that stand to this day. Mil chose to abandon further development, and focused on the Mi-26 for heavy-lift work.

At the time of writing, China has announced its intention to develop a heavy-lift helicopter to reduce reliance on outside technology and technical support for this vital element of that country's military capability.

ABOVE: **Heavy-lift helicopters can carry a variety of loads. Here, a reusable Ryan Firebee high-speed target drone is being carried by a Sikorsky HH-3.**

Anti-submarine helicopters

Fighting ships face numerous potential threats from an enemy in the form of submarine-launched torpedoes and missiles, and helicopters play a key role in detecting and neutralizing these threats. Helicopters and fixed-wing aircraft can use Magnetic Anomaly Detection (MAD) to locate submarines, and this involves the use of specialist equipment known as magnetometers. These have been developed from scientific use in geological surveys; they sense changes in the Earth's magnetic field, which can indicate the presence of large metal items under the surface of the sea, i.e. submarines. The MAD equipment is either towed by a helicopter across the surface, but can also be installed in the tail boom of fixed-wing maritime patrol aircraft. Normally, the magnetic lines of the Earth are picked up as bands, and ferro-magnetic substances cause these bands to waver and distort. This distortion is detected by MAD sensors and sent to airborne or seaborne data centres for analysis. The complex system used for the analysis of these data already has readings of known objects preloaded, so can identify most anomalies swiftly and accurately as a specific class of submarine.

In an effort to reduce the anomalies created by magnetic disturbance and thereby avoid detection, defence manufacturers have constructed hulls from non-ferrous materials such as titanium, e.g. the Russian Alpha-Class nuclear-armed submarine. However, because of items such as steel rudder surfaces and nickel alloys in the interiors of these vessels, even these supposedly less detectable boats still produce some disturbance in the water and cannot evade the anti-submarine helicopters.

"Sonar" is the acronym for SOund Navigation And Ranging, a method of underwater detection, navigation and communication that has been used since before World War II. Dipping sonar

ABOVE: **The Sikorsky SH-60 Seahawk is a development based on the UH-60 Black Hawk, and is used by many navies for anti-submarine and anti-shipping warfare, as well as Combat Search and Rescue (CSAR), Vertical Replenishment (VertRep) and MEDEVAC operations.**

LEFT: **Two Kamov Ka-25 helicopters standing on the flight deck of a Russian Navy Moskva-class cruiser.**

LEFT: **A Sikorsky MH-60R Seahawk flying a mission to evaluate Airborne Low-Frequency Sonar (ALPS) equipment.**

BELOW: **A Mark 46 Mods lightweight anti-submarine torpedo. The braking parachute slows the weapon down before entry into the sea.**

was developed by the British for the Royal Navy, and then further developed for the US Navy during the Cold War, and is primarily used for anti-submarine detection.

The dipping or dunking sonar is an instrument operated by a hydraulically powered winch on board a helicopter, on which the automatic flight control system will include a cable hover mode control. The equipment known as the sonar transducer is then lowered into the water from the helicopter. Dipping sonar can be used in either an active (sound-producing) torpedoes can use the principles of passive and active sonar to identify the target. Again, like a dipping sonar, the torpedo and then it records the echo response of the sound wave as it hits other objects and reflects back to the instrument. The response time and strength can help indicate where and what type of object the sound has encountered.

As with MAD, by comparing these readings with known readings from a very broad database, it may then be possible to identify the specific class of submarine. For example, many vessels can be identified by the sonic emissions of the power to the engines – particularly what type of energy frequency they employ. Once the area has been declared "clear", then the dipping sonar is withdrawn and recovered to the helicopter.

If a helicopter is ordered to attack a submarine, it can use torpedoes or depth charges. "Torpedo" is the common name for an underwater self-propelled missile, and homing or a passive (echo-receiving) mode. If set to active, the dipping sonar will produce a sound that is broadcast through the water, can lock on to a target by focusing on either the active sound emission or the reflection of the sound emitted and reflected back the torpedo.

Semi-active homing focuses on the last known location of a target and, once in attack range, the torpedo will "ping" for reference and locate the target. Additionally, the homing torpedo can identify other acoustic signatures beyond a ship or object; it can also recognize energy movements, such as the wake of a ship, and track accordingly.

The simplest anti-submarine weapon is the depth charge, or depth bomb, which is a helicopter-dropped large canister filled with explosive set to explode at a predetermined depth. A Nuclear Depth Bomb (NDB) is the nuclear equivalent of the conventional depth charge, and these devastating weapons have been in service with the Royal Navy, Soviet Navy and US Navy. As a nuclear warhead has much greater explosive power than a conventional depth charge, the NDB significantly increases the likelihood of destroying a submerged submarine. It was because of this much greater destructive capacity that some NDBs were designed for a variable explosive yield – from a low setting for shallow water and areas where damage had to be limited, to maximum yield for deep-water attacks. Three types of Royal Navy helicopter were cleared to carry the 272kg/600lb WE.177A nuclear depth charge: the Westland Wasp, Westland Wessex HAS.3, and Wessex HUS.

From 1966, a total of 43 NDBs were deployed aboard Royal Navy surface vessels of frigate size and larger, for use by helicopters as an anti-submarine weapon.

Soviet helicopters in Afghanistan

One of the abiding images from the television news coverage of the war in Afghanistan from 1979–89 was of Soviet helicopters in action, inserting troops, firing on enemy forces, and indeed being increasingly fired upon as the war continued.

Helicopters were by far the main element of Soviet air power used in the conflict, and it is believed that a maximum of 650 machines were deployed. Soviet helicopters were used for every possible application, but up to 300 of the force are thought to have been the Mil Mi-24 (NATO identifier Hind) gunships. Armed with machine-guns, cannon and up to 200 rocket projectiles, the Mi-24 introduced a new dimension to helicopter warfare in Afghanistan. The Mi-24 was deployed for retaliatory strikes, armed reconnaissance missions, close support of ground troops, and to identify and attack any concentration of enemy fighters. The Mil Mi-24 was a gunship and an assault helicopter which could carry eight fully equipped armed troops, but a full load did limit manoeuvrability, especially at the high altitudes experienced in Afghanistan. To save weight, cabin armour was often removed, leaving those inside vulnerable. Instead, the Mil Mi-8 helicopters would be used to transport the troops, while the Mi-24 was flown in the armed-escort role.

ABOVE: **The Mil Mi-8 (NATO identifier Hip) was widely used by the Soviet Army in Afghanistan for a number of roles, including ground attack. Some 40 were shot down by the Mujahideen during the conflict.**

RIGHT: **A Mil Mi-24 (NATO identifier Hind) of the Soviet Air Force on patrol over Afghan troops and vehicles on the Salang highway, north of Kabul, the capital of Afghanistan.**

LEFT: **Despite the variety of weaponry and troop numbers available, the invasion and ten-year occupation of Afghanistan was a Soviet failure.**

Mujahideen fighters began to use the Strela-2M heat-seeking surface-to-air missile against Soviet helicopters, and losses began to mount. The missiles were obtained from various sources, including Egypt and China, while the US Central Intelligence Agency (CIA) helped them source others. A total of 42 Soviet helicopters were shot down by various types of Strela-2 and, to overcome the problem, the Soviets began to fit exhaust shrouds to all helicopters to make them less vulnerable to the heat-seeking missile. Losses were, however, severe enough to cause a Soviet change of tactics which saw helicopters being operated at very low altitude, almost hugging the ground.

Nevertheless, for some time it was neither Soviet tanks nor ground forces that were taking the war to the Mujahideen. Mil Mi-24 helicopters were flown to attack enemy positions with guns and rockets, then insert combat troops to complete the assault on the ground. By 1986, technology was being harnessed more effectively, and the Soviet military went on the offensive against Mujahideen forces. One factor that contributed to the end of the Soviet presence in Afghanistan at that time was a US government decision to arm the Mujahideen with Stinger shoulder-fired anti-aircraft missiles. Although the guerrillas had enjoyed some success prior to the delivery of Stinger missiles using machine-guns or rocket-propelled grenades, the new missile increased the average shoot-down to one a day.

The CIA supplied at least 500 Stinger missiles to the Mujahideen from September 1986, and this enabled them to undermine a crucial Soviet advantage – the mobility and firepower provided by their helicopters. With losses mounting, the Soviet military were less inclined to risk helicopter gunships to escort the transport helicopter formations they were designated to protect, and so remained over safe territory.

On February 15, 1989, the last Soviet troops left Afghanistan as part of a general change in Soviet foreign policy.

LEFT: **With the support of the CIA, Mujahideen shoot-down successes increased, and at one point the Soviets were losing a helicopter a day. Here, rebel forces celebrate the capture of a Mil Mi-8.**

ABOVE: **Gunners in a Bell UH-1N Iroquois Huey armed with a 0.50in M213 (left) and a 7.62mm M134 Minigun (right).**

Gunships and attack helicopters

An attack helicopter can be defined as a military helicopter for which the primary role is optimized to provide close air support for troops on the ground and in the anti-tank role. The type can be armed for attacking enemy vehicles, both armoured and soft-skinned, troop concentrations, and military strongpoints such as a bunker or command post. Weapons carried can be varied, from machine-guns and cannon to unguided and guided rockets, and usually anti-tank missiles. Some of these fighting helicopters can also carry air-to-air missiles to attack or defend against enemy aircraft.

Attack helicopters evolved as a result of US Army helicopter experience from the Vietnam War, where it was realized that an agile, well-armed, purpose-designed attack helicopter was needed to counter the increasingly intense attacks from forces of the Viet Cong (VC) and North Vietnamese Army (NVA). French forces had already experimented with armed helicopters by mounting rockets and machine-guns on Piasecki H-21 Shawnee helicopters during the Algeria campaign. Basic gunships were already in service with US forces in Vietnam in the form of the Bell UH-1A. The medical version of the Bell UH-IA, the famed "Huey", entered service in South Vietnam during June 1962, but the helicopters that arrived there in September that year were heavily armed. These Utility Tactical Transport Company (UTTC) helicopters were fitted with fixed, forward-firing 7.62mm machine-guns attached on each undercarriage skid, and 2.75in rockets that could be fired simultaneously from launching tubes. These first gunships were deployed to protect CH-21 Shawnees and other transport helicopters that were used to move troops in high-risk areas. The US Army believed that they needed heavier armament, and the UH-1B had a more powerful engine, could lift more and could be additionally armed with cabin-mounted machine-guns on flexible mounts. The similarly armed UH-1C was able to offer more effective protection to the troops carried, as it had improved performance and more fuel to allow the machine to provide cover for longer periods.

LEFT: **A crewman firing a Gatling-type rotary cannon mounted on the ramp of a Sikorsky HH-3E Jolly Green Giant.**

LEFT: **The Mi-24A (NATO identifier Hind) was the first version of the heavily armed helicopter gunship which served with many Warsaw Pact nations. The type was known as "The Flying Tank" by those who flew the machine.**

However, as the fighting grew more intense, the US Army developed a requirement for a dedicated attack helicopter – this was the Advanced Aerial Fire Support System (AAFFSS) project. The winning design was the Lockheed AH-56 Cheyenne, but the US military knew it would take time to develop such a specialized machine (it never entered service due to technical problems and being over budget), but still had a military challenge growing day by day in South-east Asia. An interim solution to this gap in their inventory was required, so they invited manufacturers to propose a combat helicopter that could be in service in Vietnam relatively quickly. Designs from Sikorsky and Kaman were submitted, but it was a Bell proposal, based on the UH-1s, that was successful, and led to a contract being issued in April 1966 for 110 AH-1 Cobra attack helicopters. With tandem cockpit seating (as opposed to the typical side-by-side), the Cobra (also known as the "HueyCobra") presented a significantly smaller frontal aspect as a target, had much improved armour and an impressive performance.

In 1967, the first were deployed to South Vietnam and opened a new chapter in the history of warfare – the type was the first helicopter designed from the outset to be armed for and optimized for combat. While the Cheyenne was slowly progressing through development, the AH-1 Cobra, armed with a formidable combination of nose-mounted 20mm cannon and multiple rocket launchers or Miniguns in pod mountings, delivered a new dimension to air warfare. Later in the Vietnam War, AH-1 Cobra helicopters were deployed against ANV tanks to great success. The Lockheed AH-56 Cheyenne project was cancelled in 1972, and within weeks a new requirement was issued by the US Army that incorporated lessons learned in the Vietnam War, applying them to the ever-present threat from Soviet armour in Cold War Europe. This requirement led to the ultimate attack helicopter, the Hughes AH-64 Apache. The Soviet Union had also observed events in Vietnam, had learned from the war, and began to develop a dedicated attack helicopter.

ABOVE: **A 0.50in M213 Browning heavy machine-gun mounted in the door of a Bell UH-1 Iroquois helicopter.**

Bravo November

The aircraft known by its call sign of Bravo November is one of the most historic operated by the Royal Air Force in the post-World War II era. It is not a high-performance jet fighter or strike aircraft, but a Boeing Vertol Chinook HC2 helicopter. The RAF has flown Bravo November on operations in Northern Ireland, the Falkland Islands, Lebanon, Germany, Kurdistan, Iraq, Afghanistan and the Balkans. Four of the pilots to have flown the machine have been awarded the Distinguished Flying Cross (DFC).

The helicopter was built as a Chinook HC1 and entered RAF service in 1982. It is the only survivor of the four deployed with the British Task Force sent later that year to the Falkland Islands following the Argentine invasion. The machines were transported to the Falklands on the MV *Atlantic Conveyor*, which was sunk by an Exocet missile on May 25, 1982. Bravo November was being flown on an air test at the time, so survived and was diverted to land on HMS *Hermes*. After this lucky escape, the helicopter was in action soon after the first British troops landed on the islands, being deployed to transport 105mm artillery to Special Air Service troops on Mount Kent, where they were under fire from Argentine artillery.

Later in the Falklands campaign, the aircraft was being flown over the sea at night through a heavy snow shower, and this caused the crew to lose horizon references. The aircraft hit the sea at 185kph/115mph, but in an almost upright attitude. The pilot, Sqd Ldr Langworthy, and co-pilot, Flt Lt Lawless, successfully lifted the aircraft from the water and were, remarkably, able to fly the machine and land safely. With damaged radio equipment, however, the crew was unable to communicate with British Forces, so when approaching San Carlos, they left the aircraft's lights on in the hope that they would not be fired upon.

ABOVE: **On the Falkland Islands in 1982, Bravo November was the only Chinook to be operated during the campaign.**

ABOVE: **Despite an unplanned impact with the sea, Bravo November continued to serve on vital troop and supply missions during the Falklands campaign.**

RIGHT: **Bravo November has seen service in Northern Ireland, the Falkland Islands, Lebanon, Germany, Kurdistan, Iraq, Afghanistan and the Balkans.**

British forces had in fact heard the distress calls from the crew, so did not open fire. The machine suffered surprisingly little damage – the radio antenna was ripped off, the fuselage damaged, and a cockpit door torn off by the violence of the impact. All spares, tools and servicing manuals had been lost aboard MV *Atlantic Conveyor*, and it was two weeks before spare parts and supplies would arrive, but the ground crews kept the machine flying.

On June 2, 1982, a force of 81 paratroopers, twice the normal troop load and a record for any troop carrying helicopter, were flown from Goose Green to retake the settlement of Fitzroy. After dropping the first group, the crew returned to Goose Green to pick up a further 75 to take part in the operation. By the end of the war, Bravo November had accumulated over 100 flying hours, transported some 1,500 personnel, 558,835kg/1,232,000lb of cargo, and moved 650 prisoners and some 95 casualties. Langworthy was awarded the DFC for his actions as the pilot of Bravo November during the Falklands War.

In 2003, over than 20 years after the Falklands campaign, Bravo November was deployed during the opening night operations of Operation Desert Storm for the assault on the Al Faw peninsula, the site of a major oil refinery. Piloting the machine, Sqdr Ldr Steve Carr led the five Chinooks that were to land Royal Marine commandos at the objective.

ABOVE: **By early 2011, four pilots who had flown Bravo November on operations had each been awarded the Distinguished Flying Cross (DFC).**

During the operation, Bravo November averaged 19 flight hours a day over a three-day period, delivering combat vehicles, troops and artillery. The mission was the largest helicopter assault in UK military history, and the first helicopter assault operation since the Suez Crisis in 1956. Carr was awarded the DFC for his bravery and leadership.

The helicopter continued to be operated in Iraq, but did not have to wait another 20 years to be flown in another war. On the night of June 11, 2006, Flt Lt Craig Wilson commanded the machine on a casualty recovery mission in Helmand Province, Afghanistan. In dangerous conditions and with little experience of night flying in Afghanistan, he flew at an altitude of 46m/150ft and made a perfect approach and landing to extract the casualty. Wilson flew a similar mission just a few hours later and ran low on fuel while waiting for enemy fire to be suppressed. On landing, despite having been on duty for over 22 hours, he then volunteered to take reinforcements to deal with a deteriorating ground situation. His gallantry and extreme and persistent courage that day ensured the recovery of two badly wounded soldiers. Wilson received the DFC for “exceptional courage and outstanding airmanship” while on operations in Helmand Province.

By 2010, Bravo November was a veteran of several operational tours in Afghanistan, and January that year saw it live up to the reputation of being a lucky aircraft. After picking up six wounded soldiers following a Taliban ambush, the machine, piloted by Flt Lt Ian Fortune, came under machine-gun fire from the ground. The helicopter received several hits which damaged some systems. Fortune was also hit by a Taliban bullet that entered through the cockpit windscreen, ricocheted and shattered the visor on his flying helmet, causing him some facial injuries. After calmly reporting what had happened to his fellow crewmembers, Fortune flew the helicopter back to Camp Bastion. He was awarded the DFC for his courage. At the time of writing, Bravo November continues to be deployed in the front-line service with the RAF.

Operation Desert Storm

The Vietnam War had proved the military value of the helicopter gunship and, as that war came to an end, the US Army considered how it might apply this knowledge to the continuing Cold War situation in Europe. Attack helicopters such as the Bell AH-1Cobra were given the task of attacking, with missiles, Soviet command posts and armour behind enemy lines. Technology was also developed to help night operations and utilize superior Western technology against numerically superior Warsaw Pact forces. By the 1980s, the US had developed heavily armed helicopters as dedicated "tankbusters" to attack with guns and missiles. These tactics, although never tried in Europe, were put to the test during the first Gulf War in response to the Iraqi invasion of Kuwait, when many helicopter types saw action for the first time.

The international Coalition forces launched Operation Desert Storm on January 17, 1991, following the expiration of a deadline for Iraq to withdraw from Kuwait. Coalition naval assets operating from the Persian Gulf faced a modest but potent threat from the Iraqi Navy. Fast attack craft armed with missiles were of concern, so were attacked wherever possible. Helicopters such as the Sikorsky SH-60B Seahawk and Westland Lynx HAS Mk3, armed with Sea Skua anti-shipping missiles, together with other Coalition fixed-wing aircraft, damaged or destroyed most of the Iraqi fleet.

ABOVE: **Helicopters operated by Coalition forces were used to patrol the Persian Gulf and the Straits of Hormuz.**

The Coalition had decided to try to bomb Iraqi forces into submission before any ground war began and, as part of this strategy, US helicopter gunships such as the Bell AH-1 Cobra and Hughes AH-64 Apache were roaming over enemy positions, often at night, destroying tanks, vehicles and anti-aircraft defences. On January 18, Iraq began Scud missile attacks on Saudi Arabia and Israel, which threatened to destabilize the Coalition. As part of an all-out effort to find and destroy the Scuds on mobile launchers, special forces units were inserted by helicopter under cover of darkness deep into Iraqi territory.

LEFT: **The heavy-lift capability available to Coalition forces included CH-47 Chinooks, which were able to rapidly move supplies and equipment forward in support of advancing troops.**

ABOVE: **Although first flown operationally in the Vietnam War, the Bell UH-1 Iroquois played a key role in Operation Desert Storm.**

ABOVE: **During the Gulf War, around half of all the Sikorsky UH-60 Black Hawk helicopters in US Army service were deployed, with only two lost in combat.**

Helicopters were a vital part of the prelude to the ground war, and were used to continue attacks on Iraqi assets, virtually unmolested. When the ground war began, Apache and Cobra gunships continued to destroy Iraqi armour, and vehicles as well as bunkers. The sheer numbers of US helicopters available to fight meant that these attacks were never-ending, as wave after wave of refuelled and rearmed helicopter gunships would take their place in the assault. Apaches were even taking the surrender of Iraqi troops, who were understandably in awe of the gunship.

Most attacks by Apache helicopters were carried out from a range of some 3km/2 miles and at a height of no more than 10m/33ft to minimize the risk of a successful surface-to-air missile strike. Armed with AGM-114 Hellfire anti-tank missiles, 30mm cannon and 70mm Hydra 70 unguided rockets, the machines halted and destroyed countless Iraqi vehicle columns. Any doubts regarding the military value of helicopter gunships disappeared, as large numbers of both types decimated Iraqi armour in the open desert.

Also under cover of darkness, large numbers of Coalition troops were ferried deep inside Iraqi territory, establishing facilities for supporting the attack helicopters as well as ground troops. The fighting helicopters were complemented by general-purpose helicopters moving troops and supplies to exactly where the Coalition needed them. These included the UH-60 Black Hawk, Bell UH-1, Lynx, Gazelle, CH-47 Chinook, CH-53 Super Stallion, Puma and Super Puma. After the fighting was over, many of these helicopters were used for humanitarian missions.

LEFT: **A Sikorsky UH-60 Black Hawk of the Saudi Arabia Air Force during Operation Desert Storm.**

LEFT: **A French Army Gazelle carrying HOT missiles to attack Iraqi armour. British Army Air Corps (AAC) Gazelles were flown on scouting and observation missions.**

LEFT: **The Boeing-Sikorsky RAH-66 Comanche was an advanced armed reconnaissance and attack helicopter designed with stealth features. The development programme was cancelled in 2004.**

The anti-tank helicopter

In the late 1950s, the French had experimented with mounting wire-guided Nord SS.11 anti-tank missiles on the Alouette II (lark) helicopter, but it took the US experience in Vietnam to force the development of what was to become the anti-tank helicopter. Having developed the helicopter into potent and effective ground attack aircraft in Vietnam, the US Army then sought to apply this new expertise and the associated technology to counter the growing threat from Soviet armour in Cold War Europe. If war had broken out, and NATO faced the prospect of large numbers of Soviet tanks sweeping through the Iron Curtain, dedicated anti-tank helicopters provided one solution to this daunting military problem. During the 1970s, a considerable amount of time and money was spent on research and development of a missile-armed attack helicopter which was evolving into a primary anti-tank weapon.

While the US military had the proven Bell AH-1 Cobra, other nations were developing or were keen to acquire an anti-tank helicopter. In common with most areas of Cold War military technology, both sides would strive to develop a means of equalling, surpassing or combating each new development. The Soviet Union also began developing the helicopter as an attack aircraft; the first attempt was a heavily armed version of the Mil-8 (NATO identifier Hip) mounted with four anti-tank missiles in addition to machine-guns and rocket launchers carried on stub wings. This was followed by the formidable Mil Mi-24 (NATO identifier Hind), which also had a troop transport capability.

Helicopters are perfect for the anti-tank role because the type had been developed into a highly manoeuvrable machine that can be operated day or night to launch surprise attacks against enemy armour. A helicopter is an effective anti-tank aircraft because of the ability to be hovered behind cover, only to rise briefly and fire a missile. Early helicopter-launched missiles required a crew member to keep the target in sight and "fly" the wire-guided missile on to the target using a cockpit-mounted gun sight and a joystick controller. The development of the homing missile allowed targets to be designated by a laser so that the missile could then "home" in, although this still required the target to be kept in sight

RIGHT: **A dedicated anti-tank version of the MBB Bo105 armed with HOT missiles was procured by the German Army in the late 1970s, and a total of 212 were delivered.**

TOP: **The Lockheed AH-56 Cheyenne was developed to be the US Army's first dedicated attack helicopter, but was cancelled in 1972.** ABOVE: **The AGM-114 Hellfire anti-tank missile and 70mm Hydra 70 unguided rockets.** ABOVE RIGHT: **An AGM-114 Hellfire being loaded on the weapons pylon of an AH-64 Apache.**

from the helicopter. Continued development of even more sophisticated sensors means that helicopters such as the AH-64 Apache have only the rotor head-mounted sensor high enough too see the enemy tank before the crew can launch a missile.

In combat situations, the endurance of these missile-armed helicopters would be supported by transport helicopters to carry ground crew and armourers to refuel and reload the combat machines, getting them back into action within minutes. Forward Arming and Refuelling Points (FARP) are set up at pre-arranged locations for the replenishment process, which can take place with engines running and rotors turning.

Helicopters have been used in anti-tank combat in many conflicts. During the 1973 Yom Kippur War, Israel had used fast fixed-wing jet aircraft against Syrian and Egyptian tanks and, although victorious, they were aware of the disadvantages compared to helicopters. As a result, Israel began procuring a force of anti-tank helicopters, opting for the Bell AH-1 Cobra armed with 20mm cannon and anti-tank weapons, and also the Hughes 500MD Defender fitted with a nose-mounted sight and the capability to carry usually two Tube-launched Optically-tracked Wire-guided (TOW) missiles. In 1982 during the Lebanon War, these helicopters were used against Syrian armour, and destroyed many of the then modern, Soviet-designed T-72 tanks. This was of great interest to NATO strategists, who would hope to replicate the success if the Cold War in Europe ever became hot. The anti-tank helicopter was proved to be a vital element of any armed forces facing the potential threat of an attack by enemy armour.

BELOW: **The 9K11 Malyutka (NATO identifier AT-3 Sagger), a wire-guided anti-tank guided missile, was developed in the Soviet Union to be both man-portable and helicopter-fired.**

AgustaWestland AH-64W Apache Longbow

On November 15, 1972, the US Army issued a Request For Proposal (RFP) for an Advance Attack Helicopter (AAH), and specified a low-flying, highly manoeuvrable machine. Hughes produced the (Model 77) YHA-64A and Bell the (Model 409) YHA-63A. After evaluation, the US Army selected the Hughes machine. Bell used development data from the YHA-63A to produce the AH-1 Cobra gunship, and the AH-64A entered service in January 1984. Westland Helicopters built 67 of the type as the WAH-64A Apache for the British Army Air Corps (AAC), and it remains in service as the Apache Longbow.

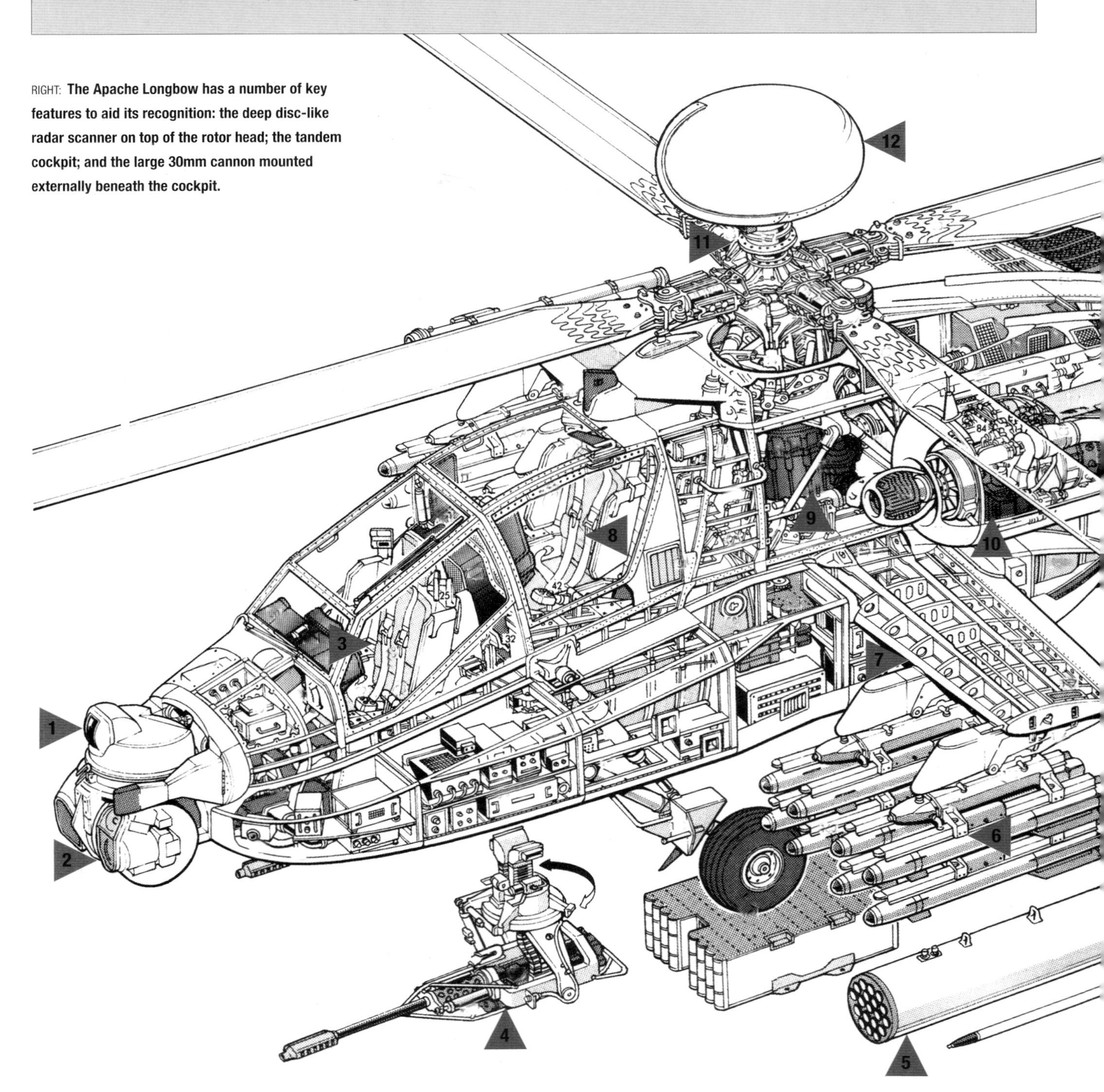

RIGHT: **The Apache Longbow has a number of key features to aid its recognition: the deep disc-like radar scanner on top of the rotor head; the tandem cockpit; and the large 30mm cannon mounted externally beneath the cockpit.**

LEFT: **The Longbow mast-mounted fire-control radar detects, classifies, prioritizes (up to 128 in a minute) and engages targets in all conditions – whether those targets are stationary or mobile, multiple on the ground or airborne.**

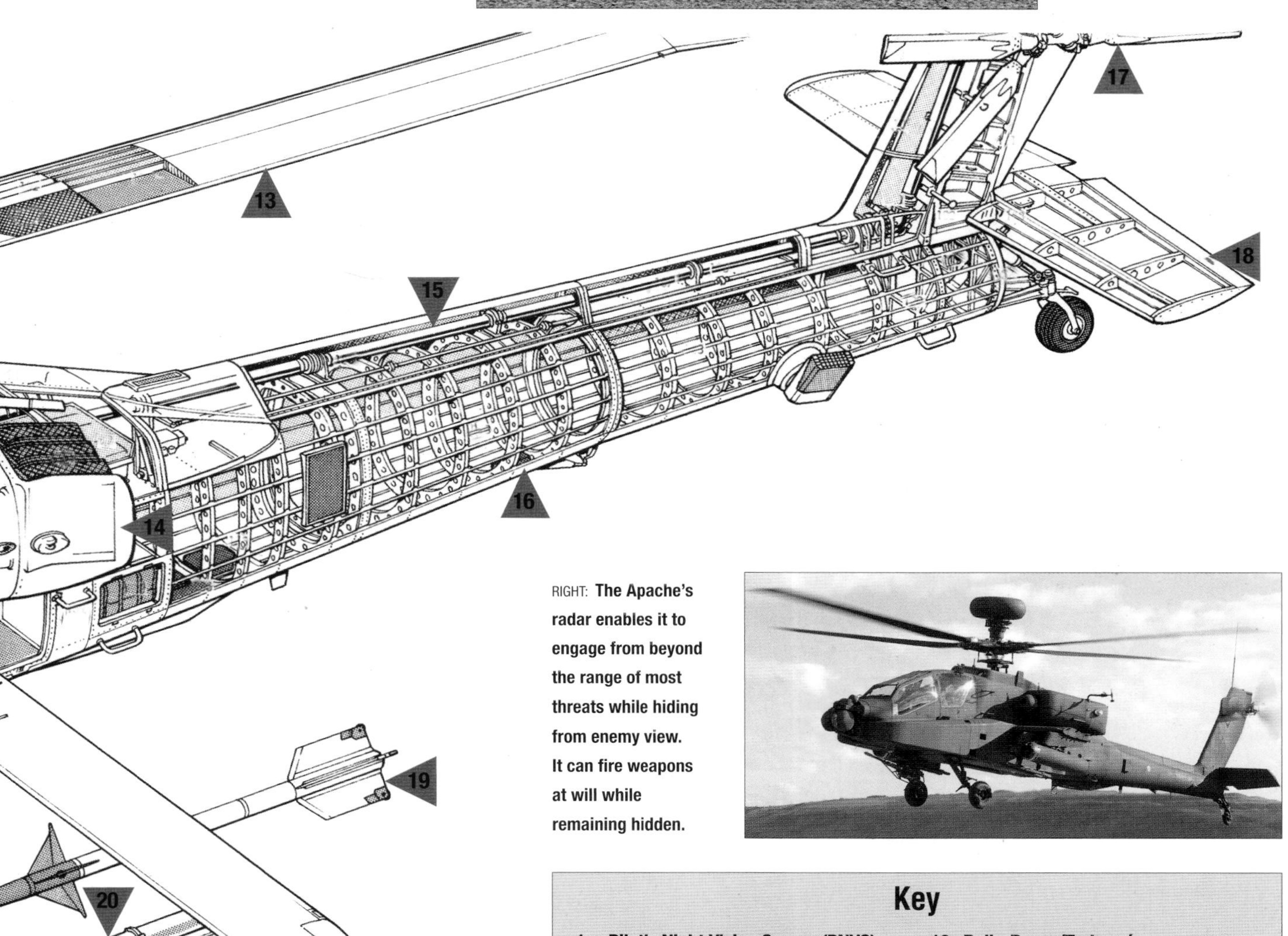

RIGHT: **The Apache's radar enables it to engage from beyond the range of most threats while hiding from enemy view. It can fire weapons at will while remaining hidden.**

Key

1. Pilot's Night Vision Sensor (PNVS)
2. Target Acquisition and Designator Sight (TADS)
3. Co-pilot/gunner's seat
4. 30mm M230 Chain Gun cannon and mounting
5. 70mm air to-ground rocket and pod
6. AGM-114 Hellfire missile
7. Stub wing
8. Pilot's seat
9. Main gearbox
10. Rolls-Royce/Turboméca RTM322 turboshaft engine
11. Main rotor head
12. Longbow radar
13. Rotor blade
14. Infra-red suppressor duct
15. Tail rotor driveshaft
16. Tail boom structure
17. Tail rotor
18. Tailplane
19. AIM-9L Sidewinder missile
20. AIM-92A Stinger missile

LEFT: **The first type chosen for presidential service was the Bell H-13 Sioux, but this was soon replaced by a larger and more comfortable machine, the Sikorsky HUS-1 Seahorse, operated by the US Marine Corps. The type was later used by President John F. Kennedy. Note the machine is fitted with inflatable flotation bags on the main wheels for an emergency landing on water.**

US presidential helicopters

In 1957, Dwight D. Eisenhower became the first serving US President to use a helicopter for his official duties. While the US military had been using helicopters since 1944, the cautious Secret Service had resisted their introduction as a presidential transport due to safety concerns over what was perceived to be new technology. However, in Cold War 1956, with the potential for Armageddon to be unleashed with just a few minutes' notice, it was clear that a quick means of evacuating the President was required. The need for a presidential helicopter was agreed, and the search for appropriate candidate types began. Safety and reliability were high on the list of deciding factors, and the first chosen was the Bell H-13 Sioux, proven in service during the Korean War. In addition to the pilot, the small helicopter could only carry the President and one other passenger, a Secret Service protection agent. Bell modified an H-13J specifically for this usage, and it featured all-metal rotor blades, and special arm and foot rests for the right-hand seat, covered with custom upholstery. A frameless Plexiglas nose was tinted dark blue to reduce heat and glare. The H-13 was considered an odd choice by many, not least because of its size. It had a range of only 242km/150 miles, it was comparatively slow, and it had just one pilot, while larger machines had two in case one become incapacitated. The presidential H-13J, serial number 57-2729, also had a rotor-brake to reduce the shutdown time and allow the President a swift exit. Eisenhower's first flight left the South Lawn of the White House on July 12, 1957. In September 1957, he took a flight in a Sikorsky HUS-1 of the US Marine Corps, soon appreciated the shortcomings of the small Bell machine,

LEFT: **In 1961, the Sikorsky VS-34 was replaced in service with the Sikorsky H-3 Sea King, a larger and more comfortable machine operated by the USMC as the VH-3D. All presidential helicopters use the call-sign "Marine One".**

RIGHT: **The Sikorsky VH-3D Sea King was supplemented in service by the Sikorsky CH-53 Super Stallion. All presidential flights are operated by Marine Helicopter Squadron One (HMX-1).**

and requested that the Sikorsky machine became the presidential helicopter. Until then, the President had always been flown by the US Air Force, but as the Sikorsky was not in USAF service, the honour now fell to both the United States Marine Corps and the US Army, who took it in turns to transport the Commander-in-Chief. This split responsibility remained, albeit with different helicopters, right through to 1976, when the administration of Gerald Ford removed presidential helicopter obligations from the US Army for economic reasons.

Most people are familiar with "Air Force One", the Presidential Boeing 747 operated by the USAF. However, the President's journey to Air Force One often begins with a flight from the White House lawn in a helicopter always known as "Marine One", the call sign of any one of the fleet of USMC machines available to carry the President. When the Army operated presidential helicopters, the call sign was, logically, "Army One". In 1961, the HUS-1 was replaced by another Sikorsky machine, the VH-3A VIP version of the Sea King. The VH-3D replaced some VH-3As in 1978. The Sikorsky VH-60 was added to the Marine One fleet of helicopters from 1989, in both VH-60D Nighthawk and VH-60N Whitehawk variants. Marine One is typically operated by the USMC HMX-1 Nighthawks squadron, and is currently either the VH-3D Sea King or the VH-60N Whitehawk, which are both due to be replaced within the next decade.

Over 800 USMC ground crew and pilots are involved in the operation of the Marine One fleet, which is based in Quantico, Virginia. It is notable that marine aviators flying Marine One do not wear regular flight suits during presidential flights, but wear dress uniform. As a security measure, Marine One always flies in a group of up to five identical helicopters. One of these carries the President, while the others are decoys to confuse any potential attacker on the ground, and change formation to add to the confusion.

LEFT: **The Sikorsky VH-60D Nighthawk version of the SH-60 Seahawk entered service with HMX-1 in 1989. The Sikorsky VH-60N Whitehawk variant is also in Marine One service. Along with all the VH-3D helicopters, the type is due for replacement within the next decade.**

LEFT: **US Marine Corps troops being flown to the beachhead in Boeing Vertol CH-46 Sea Knight helicopters. A Landing Craft, Air Cushion (LCAC) is approaching the beach.**

Helicopter assault ships

The development of the military helicopter significantly increased the options open to naval planners in terms of how they could conduct operations. The Allies' and particularly the US experience in World War II in the Pacific demonstrated how air power could turn battles, provide vital top cover for seaborne assaults, and attack defensive strongpoints as the first part of a seaborne assault. Nevertheless, there were still many restrictions on the type of coastline that could be considered for amphibious assault, including tides and the steepness of the beach, which could rule out the use of landing craft. These factors limited military options but the arrival of the helicopter made it possible, in theory, to land troops anywhere, regardless of tides or beach topography, and pose a threat that is hard to defend against.

Today's assault ships are designed to support amphibious warfare – the landing and support of troops on to or near enemy territory by an amphibious assault. They have their origins in the ships of the 1950s that were used by some navies in early airborne assaults, such as by the Royal Navy in Suez. Ships were then converted to carry more, or perhaps only, helicopters, and this evolution led to purpose-designed assault ships that can not only accommodate helicopters but also amphibious landing craft, by the inclusion of a well deck. These ships are quite different from a fixed-wing aircraft carrier, but do also have the ability to operate V/STOL aircraft.

ABOVE: **HMS *Ocean* (L12), a Royal Navy Landing Ship, Personnel, Helicopter (LPH) carrying Westland Sea King helicopters and a Boeing CH-47 Chinook.**
RIGHT: **HMS *Albion* was built as a Centaur-class aircraft carrier in 1947, and converted to a Commando Carrier in 1962.**

LEFT: **USS *Boxer* (LHD-4) is a Wasp-class amphibious assault ship. The vessel can carry up to 42 aircraft, including the Sikorsky CH-53 Super Stallion and the Bell-Boeing V-22 Osprey.**

The largest fleet of this type of ship is operated by the US Navy, and includes the Tarawa-class vessel first deployed in the 1970s and the Wasp class which entered the US fleet in 1989. These US Navy ships are specifically known as amphibious assault ships, whereas other navies tend to refer to the type as amphibious warfare ships.

The Wasp class carries a variety of assault helicopters, plus six to eight fixed-wing AV-8B Harrier aircraft for close air support. The helicopters carried on board a single ship can include twelve CH-46 Sea Knights, four CH-53E Sea Stallions, four AH-1W Super Cobras and three UH-1 Iroquois. A US Marine Corps Battalion of up to 1,900 troops can be embarked, as well as trucks, assault vehicles and tanks.

The first Royal Navy ship to be built specifically for the amphibious assault role was HMS *Ocean* (L12), which entered service in 1998. It can carry up to 12 Sea King or four Merlin helicopters, six Lynx or AH-64 Apache helicopters. and a full battalion of troops, as well as light vehicles and equipment. A large flooding dock allows the ship to carry and deploy landing craft and large assault hovercraft.

By way of comparison, France's Mistral-class amphibious assault ships can transport and deploy 16 NI Industries NH90 or Eurocopter Tigre/Tiger helicopters, four landing craft, up to 70 vehicles, including 13 tanks, or an entire 40-strong tank battalion and 450 troops. The flight deck has six helicopter landing spots, one of which is capable of supporting a heavy helicopter. The hangar deck can hold 16 helicopters, and is even equipped with an overhead crane – two aircraft lifts connect the hangar to the flight deck. Every helicopter operated by the French military is capable of flying from Mistral-class ships.

BELOW: **A US Navy SH-60F Seahawk, assigned to Helicopter Anti-Submarine Squadron 14 (HS-14), flying over the Republic of Korea (RoK) Dokdo-class amphibious assault ship *Dokdo* during joint exercises in 2010.**

Tilt-wing and tilt-rotor

The path that led to the Bell-Boeing V-22 Osprey in service today with the US Air Force and US Marine Corps was beaten by a relatively small number (in aerospace industry terms) of gifted, forward-thinking and stubborn designers who refused to accept that a viable tilt-rotor or tilt-wing aircraft could not be designed. The benefit of both designs is that they combine the take-off, landing and hover capability of a helicopter with the range and performance of a fixed-wing aircraft, achieved through tilting the wing with engines and rotors attached (tilt-wing) or just the rotors (there were usually more than one) through around 90 degrees (tilt-rotor).

The dream of lifting off from a city centre helipad and then flying at high speeds was of great interest to the airline industry. Some individuals among the military, having realized the limitations inherent in a helicopter, could also see the remarkable versatility and capability this type of machine could give the armed forces – landing and taking off virtually anywhere, and transiting between locations at high speed. There were, however, enormous technical problems to overcome, as well as numerous design failures and mishaps. A number of the designs that added to an understanding of the capability warrant examination.

Credit for creating the first working tilt-rotor aircraft goes to the team at the splendidly named Transcendental Aircraft Corporation of Pennsylvania. In 1954 and 1955, their experimental aircraft were routinely achieving up to 70 degrees of tilt, with the wings supporting over 90 per cent of the aircraft's weight.

RIGHT: **The Canadair CL-84 Dynavert was a V/STOL turbine-powered tilt-wing aircraft, four examples of which were manufactured by Canadair between 1964 and 1972. Although two CL-84s crashed due to mechanical failure, the type was considered a successful experimental type.**

ABOVE: **The Vertol VZ-2 was built in 1957 to investigate tilt-wing technology for Vertical Take-Off and Landing (VTOL). The T-tail incorporated small ducted fans to act as thrusters for greater control at low speeds. On July 23,1958, the aircraft made a full transition from vertical to horizontal flight. This groundbreaking research aircraft is now preserved by the National Air and Space Museum.**

Bell had been working on the tilt-rotor concept since the late 1940s, and built the XV-3. A slender metal wing was mounted mid-fuselage, and a large 7.63m/25ft helicopter-type rotor was mounted on each wing tip. The machine was powered by a single Pratt & Whitney R-985-AN-1 Wasp Junior radial piston engine. Power to the rotors was transmitted via gearboxes and drive shafts. Wind-tunnel testing showed positive results, but what did not exist at this point in aerospace history was the ability to predict in advance the impact of what is known as aeroelasticity – the effect that the large, slow-turning rotors would have on the wing structure and other flexible components of the aircraft, which can be both destabilizing and damaging to the structure. The XV-3 was plagued by the resulting instabilities, which ultimately caused the first prototype to crash. The second prototype which, like the

LEFT: **The Ling-Temco-Vought (LTV) XC-142 was designed to investigate the operation of V/STOL transports. The first conventional flight took place on September 29, 1964, and the first transitional flight on January 11, 1965.**

first prototype, was modified and refined after each series of testing, finally achieved a full 90-degree conversion to forward flight in December 1958, three years after the first hover flight. Although the XV-3 also successfully demonstrated the feasibility of the tilt-rotor concept, it was not taken any further, but provided a wealth of data for subsequent related research programmes.

Meanwhile, Vertol were working on tilt-wing concepts, and produced the VZ-2A. A single Lycoming turboshaft engine was mounted above the fuselage, driving two wing-mounted propellers for lift as well as two ducted fans, one in the fin and the other in the horizontal stabilizer. These fans were used for pitch and yaw control of the craft during hovering and transition flight. The VZ-2A was flown vertically in April 1957. On July 15, 1958, the first complete transition took place, demonstrating vertical take-off to forward flight and then back to a vertical landing, proving the tilt-wing concept.

Hiller's large (for a test vehicle) X-18 of 1959 mated the fuselage of a Chase transport aircraft with a high-set tilting wing, but the design for the US Air Force did not progress beyond 20 test flights. In 1962, Kaman developed a tilt-wing aircraft for the US Navy which had a Grumman Goose fuselage mated to an all-new Kaman-designed tilting wing. This exotic design did not progress beyond the wind-tunnel testing.

The Vought/Hiller/Ryan XC-142 tilt-wing aircraft was developed with US Government backing to create a design that, unlike most preceding tilt-wing experiments, really could have tri-service military applications for the US Army, USN and USAF. Proposed as a transport aircraft, the XC-142 was 18m/59ft long and had an all-up weight of 16,900kg/37,258lb. It was powered by four General Electric T64-GE-1 turboprop engines mounted on the wings, cross-linked to drive four-bladed propellers. The engines also drove a fifth propeller – a three-bladed unit in the tail, which rotated in the horizontal plane. Despite great promise, shortcomings and a number of accidents halted progress but, again, much useful data was gathered for later programmes.

ABOVE: **The Bell XV-15 tilt-rotor aircraft, which first flew in May 1977, was the first of its kind to demonstrate the high-speed performance of the type compared to that of a conventional helicopter. It had a top speed of 557kph/345mph.** BELOW: **The Bell XV-15 undergoing a test flight programme at Ames Research Centre (ARC), located at Moffett Field in Silicon Valley, California.**

LEFT: **A Northrop Grumman RQ-8A Fire Scout approaching for the first autonomous landing aboard USS *Nashville* (LPD-13) during sea trials in 2006.**

Vertical Unmanned Aircraft Systems (VUAS)

Unmanned rotary aircraft have been in military service since the early 1960s, and numerous projects currently under development will see these drones entering service in increasing numbers over the next decade.

The Gyrodyne QH-50 DASH (Drone Anti-Submarine Helicopter) was a small, coaxial rotor helicopter built as a long-range anti-submarine weapon to be deployed on US Navy warships too small to operate a full-sized ASW helicopter. In the late 1950s, the US had to urgently increase anti-submarine capability to counter the threat of the growing numbers of Soviet submarines, and the QH-50 gave an anti-provide unprecedented situational awareness and precision other duties. The machine was powered by a Boeing T50-4 turboshaft engine and could carry two Mk 44 torpedoes or a nuclear depth charge. A total of 378 were produced before production ended in January 1966. The QH-50 DASH required two controllers – one on the flightdeck and another in the Combat Information Centre (CIC). The flight-deck controller would handle take-off and landing, while the CIC controller improve this far-from-ideal situation, television cameras were controls and radar. The type had a range of up to 35km/22 miles, so the submarine would have no warning of the attack until a torpedo had entered the water. The CIC controller could not see the aircraft and occasionally lost control, so to flew the drone to the target and attacked using semi-automated installed experimentally late in the programme. The DASH programme was cancelled in 1969, but some modified machines were operated in a number of roles during the Vietnam War. The Japanese Maritime Self-Defense Force (JMSDF) also operated a fleet of 20 of the type until 1977.

The Northrop Grumman Corporation Fire Scout, a Vertical Unmanned Aircraft System (VUAS), is under development to submarine capability to existing ships that were outdated for targeting support for the US military. The MQ-8B Fire Scout is based on the Schweizer Model 333 helicopter, a proven design. The machine can take off and land autonomously on any air-capable warship or at unprepared landing zones in proximity to the edge of the forward battle area.

With vehicle endurance greater than eight hours, the MQ-8B Fire Scout will be capable of operating to provide coverage some 161km/100 miles from the launch site. The type is equipped with electro-optical/infra-red sensors and a laser pointer/laser rangefinder to enable the machine to locate, track and designate targets.

In January 2006, two Fire Scout VUAS drones completed nine autonomous ship-board landings on USS *Nashville* (LPD-13), the first time a VUAS in USN service performed vertical landings on a moving ship without a pilot controlling the aircraft.

Lockheed Martin and the Kaman Aerospace Corporation have successfully transformed the proven Kaman-built K-MAX power-lift helicopter into an unmanned aircraft for autonomous or remote-controlled cargo delivery. This unmanned helicopter can routinely be used to undertake

ABOVE: **The Bell/Textron Eagle Eye in service with the US Coast Guard can fly at over 370kph/230mph, has a range of 1,481km/920 miles, and can fly at up to 6,096m/20,000ft for some six hours.** RIGHT: **In April 2011, US Navy Fire Scouts were despatched to support US Army and Coalition forces in Afghanistan. Fire Scout is a small helicopter able to stay aloft for more than eight hours at altitudes up to 5,182m/17,000ft.**

battlefield cargo resupply missions in hazardous conditions, and has drawn considerable interest from the US Marine Corps aircraft, who are at present evaluating the type. The aircraft would enable the USMC to deliver supplies, day or night, to precise locations without the risk of losing aircrew. The aircraft can fly at higher altitudes with a larger payload than any other rotary-wing unmanned vehicle, and is fitted with a four-hook carousel. The K-MAX drone can also deliver more cargo to more locations in one flight. The machine can lift 2,725kg/6,000lb at sea level, and more than 1,817kg/4,000lb at an altitude of 4,575m/15,000ft.

Meanwhile, Northrop Grumman are developing the Fire-X, which the company describes as an "expanded capability vertical unmanned aircraft system" that combines the reconnaissance, surveillance and target acquisition equipment fitted in the MQ-8B Fire Scout, with the extended range, payload and cargo-lifting capabilities of the civilian Bell Model 407 helicopter. The Fire-X is a fully autonomous, single-engine unmanned helicopter that can carry an array of battlefield intelligence, surveillance and reconnaissance sensors to meet increased demands for enhanced situational awareness. The type can lift up to 1,363kg/3,000lb in total, of which 1,181kg/2,600lb is carried externally. The Fire-X has a flight endurance of up to 14 hours.

Unmanned helicopters are offering the military an increased, cost-effective capability and are expected to become, like fixed-wing drones, a major part of the air assets available to military commanders in the field.

LEFT: **Northrop Grumman state that: "The transformational Fire Scout Vertical Take-off and Landing Tactical Unmanned Aerial Vehicle system provides unprecedented situation awareness and precision targeting support for US armed forces of the future."**

Glossary

AAC Army Air Corps (UK).
AAF Army Air Forces (US).
AAM Air-to-Air Missile.
aerodynamics Study of how gases, including air, flow and how forces act upon objects moving through air.
AEW Airborne Early Warning.
ailerons Control surfaces at the trailing edge of each wing used to make the aircraft roll.
angle of attack Angle of a wing or rotor blade to the oncoming airflow.
anti-torque To counter the effect of torque, often applied to a system.
ASDIC Anti-Submarine Detection Investigation Committee; a system that uses pulses of sound to detect objects underwater, invented during World War I and refined during World War II.
ASM Anti-Ship Missile.
ASR Air Sea Rescue; *see also* SAR.
ASW Anti-Submarine Warfare.
autorotation The movement of relative wind up through the rotor blades, causing them to turn with enough speed to generate lift and carry the aircraft aloft without an engine.
AWACS Airborne Warning and Control System.
blister A streamlined, often clear, large fairing on an aircraft body, housing guns or electronics.
BVR Beyond Visual Range.
CASEVAC Casualty Evacuation.
ceiling The maximum height at which an aircraft can operate.
coaxial Contra-rotating superimposed rotors which cancel out any torque effect.
collective/collective control Essentially the "up" or "down" control that changes, collectively, the angle of all rotor blades simultaneously, and increases or decreases the lift that the rotors provide to the aircraft, allowing the helicopter to gain or lose altitude.
control tubes Push/pull tubes that change the pitch of the rotor blades.
CSAR Combat, Search and Rescue.
cyclic Cyclic control changes the angle of attack of the main rotors unevenly – on one side of the helicopter the angle of attack (and therefore lift) is greater.
dihedral The upward angle of the wing formed where the wings connect to the fuselage.
dipping sonar Sonar that is lowered into the sea by a helicopter to listen for submarines beneath the surface.
dorsal Pertaining to the upper side of an aircraft.

BELOW: **The Sikorsky (S-49) R-6A was developed from the R-4, but in order to provide improved performance, the fuselage was completely redesigned. The R-6A was fitted with the rotor and gearbox from the R-4 and was powered by a Franklin 0-405-9 radial piston engine.**

drag The force that resists the motion of the aircraft through the air.
drone An unmanned aircraft controlled by radio or other means.
dynamic components The main rotating part of a helicopter airframe.
ECM Electronic Countermeasures.
elevators Control surfaces on the horizontal part of the tail that are used to alter the aircraft's pitch.
ELINT Electronic Intelligence.
eshp Equivalent shaft horsepower.
FBW Fly-By-Wire.
fin The vertical portion of the tail.
flaps Moveable parts of the trailing edge of a wing used to increase lift at slower air speeds.
FLIR Forward Looking Infra-Red.
g The force of gravity.
hp Horsepower.
hub This sits atop the mast, and connects the rotor blades to the control tubes.
HUD Head-Up Display.
jet engine An engine that works by creating a high-velocity jet of air to propel the engine forward.
JMSDF Japanese Maritime Self-Defense Force.
LCD Liquid Crystal Display.
leading edge The front edge of a wing or tailplane.
MAD Magnetic Anomaly Detection; a technique for locating submarines by detecting their metal mass.
mast Rotating shaft from the transmission, connecting the rotor blades to the helicopter.
MEDEVAC Medical Evacuation.
NATO North Atlantic Treaty Organization.
nautical mile 1.852km/1.1508 miles.
pitch Rotational motion in which an aircraft turns around its lateral axis. Alternatively, increased or decreased angle of the rotor blades to raise, lower or change the direction of the rotors' thrust force.
port Left side.
RAAF Royal Australian Air Force.
radome Protective covering for radar, made from material through which radar beams can pass.
RAF Royal Air Force.
RCAF Royal Canadian Air Force.
RN Royal Navy.
roll Rotational motion in which the aircraft turns around its longitudinal axis.
root The inner end of the blade where the rotors connect to the blade grips.
rotor The rotary wing formed of spinning blades that act as a wing to generate lift.
rudder The parts of the tail surfaces that control yaw (left and right turning).
SAAF South African Air Force.
SAM Surface-to-Air Missile.
SAR Sea Air Rescue/Search and Rescue.
shp Shaft horsepower.
SLR Side-Looking Airborne Radar.
sonar Acronym for Sound Navigation and Ranging, a technique that uses sound propagation, usually underwater, to detect other vessels.
sponson An aerodynamic fairing on the lower side of an helicopter, often housing fuel.

ABOVE: **The Aérospatiale Gazelle is a light multi-role helicopter that has been in military service since the early 1970s. The tail rotor housed within the tail boom is one of its key recognition features.**

starboard Right side.
STOL Short Take-Off and Landing.
supersonic Indicating motion faster than the speed of sound.
tailplane Horizontal part of the tail, known as horizontal stabilizer in North America.
thrust Force produced by an engine, pushing the aircraft forward.
torque The tendency of an engine and the aircraft it is mounted in to spin around in the opposite direction, a rotor being driven by the engine.
UHF Ultra High Frequency.
USAAC United States Army Air Corps.
USAAF United States Army Air Forces.
USAF United States Air Force.
USCG United States Coast Guard.
USMC United States Marine Corps.
USN United States Navy.
ventral Pertaining to the underside of an aircraft.
VHF Very High Frequency.
V/STOL Vertical/Short Take-Off and Landing.

Index

LEFT: **Bell UH-1 Iroquois.**